W0050296

An Introduction to Ionic Liquids

An Introduction to Ionic Liquids

Michael Freemantle

RSCPublishing

ISBN: 978-1-84755-161-0

A catalogue record for this book is available from the British Library

© Michael Freemantle 2010

All rights reserved

Apart from fair dealing for the purposes of research for non-commercial purposes or for private study, criticism or review, as permitted under the Copyright, Designs and Patents Act 1988 and the Copyright and Related Rights Regulations 2003, this publication may not be reproduced, stored or transmitted, in any form or by any means, without the prior permission in writing of The Royal Society of Chemistry or the copyright owner, or in the case of reproduction in accordance with the terms of licences issued by the Copyright Licensing Agency in the UK, or in accordance with the terms of the licences issued by the appropriate Reproduction Rights Organization outside the UK. Enquiries concerning reproduction outside the terms stated here should be sent to The Royal Society of Chemistry at the address printed on this page.

The RSC is not responsible for individual opinions expressed in this work.

Published by The Royal Society of Chemistry,
Thomas Graham House, Science Park, Milton Road,
Cambridge CB4 0WF, UK

Registered Charity Number 207890

For further information see our website at www.rsc.org

Preface

The 30 March 1998 issue of *Chemical & Engineering News* (C&EN), the weekly news magazine of the American Chemical Society, published an article entitled "Designer Solvents: Ionic Liquids May Boost Clean Technology Development." The article described research on and potential applications of room-temperature ionic liquids. These materials are salts, typically with large organic cations and inorganic anions, that are liquid at ambient temperatures.

Until the mid-1990s, research on room-temperature ionic liquids had been a quiet backwater of chemistry. Relatively few scientists were familiar with the topic. I had certainly never heard of them.

During the 1980s and early 1990s, no more than 25 scientific papers were published on the topic each year. Much of this research focused on the use of ionic liquids for electrochemical applications and as solvents for a small number of organic reactions such as Friedel–Crafts acylations and alkylations.

The C&EN article caused quite a stir and, so I am told, sparked an explosion of research activity in the field. Chemists who were not familiar with these liquids soon appreciated that the cations and/or anions could be readily changed to optimize control over both yield and selectivity in a chemical reaction.

The possible use of these liquids for green chemistry was particularly exciting. Ionic liquids dissolved almost anything and could therefore potentially be used as environmentally-safe solvents for a whole range of chemical reactions. The few chemists that were carrying out research on these liquids in the 1990s hoped they would replace the volatile organic

An Introduction to Ionic Liquids
By Michael Freemantle
© Michael Freemantle 2010
Published by the Royal Society of Chemistry, www.rsc.org

solvents that are used in the chemical industry to make pharmaceuticals and other products. These organic solvents are generally toxic. Many are flammable. In addition, because they are volatile, there is always a risk that they will escape into the atmosphere and pollute the environment. Ionic liquids, in contrast, have no measurable vapour pressure, which means that they do not release noxious gases into the atmosphere.

Ionic liquids also act as catalysts for a range of chemical reactions. They are generally immiscible with non-polar organic liquids, and can therefore be used for separations and readily recycled. Above all, by ringing the changes on the cations and anions, the chemical, physical and biological properties of ionic liquids can be designed to suit various applications, not only as solvents and catalysts but also as other types of useful materials.

A decade after the publication of the C&EN article, the number of research publications on ionic liquids was running at around 2000 a year. The growth had been almost exponential. By August 2007, over 8000 publications, including 900 patents, on ionic liquids had appeared, 97% of which were published following the 1998 C&EN article, according to Robin Rogers, chemistry professor at Queen's University of Belfast (QUB) and Director of the Center for Green Manufacturing at the University of Alabama, Tuscaloosa, USA. I was the author of that article. At that time I was European Science Editor of C&EN.

This book provides an introduction to ionic liquids, their chemical, physical and biological properties, and their wide-ranging uses and potential applications. The science is illustrated by a representative selection of examples of research chosen principally from the primary research literature.

The book aims to help students and teachers in schools, colleges and universities and also scientists in industry and government to familiarize themselves with one of the most rapidly advancing and exciting fields of science and technology today.

My fascination with ionic liquids started in 1997 during my research for the 1998 C&EN article. Since then, I have written over 50 articles and news items on the topic. I would like to express my gratitude to the researchers at the Queen's University Ionic Liquids Laboratories (QUILL) Research Centre in Belfast, Northern Ireland who have helped me with these articles and news items by keeping me up-to-date with the latest information on ionic liquids research. In particular, I would like to thank QUILL director and Queen's University of Belfast (QUB) chemistry professor Ken Seddon, who introduced me to ionic liquids in 1997, and Robin Rogers for many stimulating discussions on the topic.

I am also grateful to Janet Freshwater and her colleagues at the Royal Society of Chemistry for their help in the production of this book, and to my wife, Mary, for her support.

Finally, I would like to dedicate this book to my grandchildren, Jack, Sophia, Eloise and Joel. Who knows, one day one of them may decide to pursue a career in chemistry.

Michael Freemantle

Contents

An Introduction to Ionic Liquids
By Michael Freemantle
© Michael Freemantle 2010
Published by the Royal Society of Chemistry, www.rsc.org

CHAPTER 1

Introduction

1.1 DEFINITION OF IONIC LIQUIDS

Ionic liquids are liquids that consist exclusively or almost exclusively of ions. They therefore exhibit ionic conductivity. This definition includes liquids that are traditionally known as molten salts or fused salts, which have high melting points.

Over the past two decades or so, the use of the term "ionic liquids" has generally been limited to those liquids, as defined above, that have melting points or glass-transition temperatures below 100 °C. They are typically organic salts or eutectic mixtures of an organic salt and an inorganic salt.

Aqueous solutions of salts are not classified as ionic liquids as they do not consist exclusively of ions.

Liquids, such as ethylammonium nitrate, that consist almost exclusively of cations and anions with only a small amount of molecular species are classified as ionic liquids (see protic ionic liquids, Section 1.6).

Binary mixtures that are liquid and consist entirely of ions, such as the haloaluminate systems (Section 1.7), are also known as ionic liquids.

Ionic liquids that are liquid at or around room temperature are often called "room-temperature ionic liquids."

1.2 SYNONYMS

Following their discovery in the late nineteenth century and early twentieth century (Chapter 2), various synonyms and abbreviations

An Introduction to Ionic Liquids
By Michael Freemantle
© Michael Freemantle 2010
Published by the Royal Society of Chemistry, www.rsc.org

have been and are used in the scientific literature for organic salts with low melting points or low glass-transition temperatures:

- ionic liquid (IL)
- room-temperature ionic liquid (RTIL)
- ambient-temperature ionic liquid
- non-aqueous ionic liquid (NAIL)
- molten organic salt
- fused organic salt
- low melting salt
- neoteric solvent
- designer solvent.

1.3   ATTRACTION OF IONIC LIQUIDS

Since the late 1990s, ionic liquids have attracted the attention of chemists around the world for various reasons:

- Ionic liquids have opened up a new face of chemistry. Before 1998, there were relatively few studies of chemistry at temperatures below $100\,°C$ in a liquid environment that was entirely ionic compared with chemistry in a molecular environment.
- The scientific potential for research on ionic liquids is virtually unlimited. To date more than 1500 ionic liquids have been reported in the scientific literature. In theory at least, a million or so simple ionic liquids are possible. An almost limitless number of ionic liquid systems are theoretically possible by mixing two or more simple ionic liquids.
- Unlike organic molecular solvents, ionic liquids have negligible vapour pressures, and therefore do not evaporate under normal conditions.
- Ionic liquids are generally non-flammable and many remain thermally stable at temperatures higher than conventional organic molecular solvents.
- Ionic liquids have wider liquid ranges than molecular solvents.
- Ionic liquids have a wide range of solubilities and miscibilities. For example, some ionic liquids are hydrophilic while others are hydrophobic.
- Ionic liquids have wide electrochemical windows.
- Ionic liquids can be used as reaction media and/or catalysts for a wide variety of chemical reactions.

- Ionic liquids can be used for separations and extractions of chemicals from aqueous and molecular organic solvents.
- Ionic liquids can be readily recycled following use as solvents and/or catalysts.
- The physical, chemical and biological properties of ionic liquids can be "tuned" or "tailored" by:

  (a) switching anions or cations;
  (b) by designing specific functionalities into the cations and/or anions;
  (c) by mixing two or more simple ionic liquids.

- Because ionic liquids consist of cations and anions, they have dual functionality. They therefore impart a unique architectural platform compared with molecular liquids. Consequently, ionic liquids can potentially be exploited as solvents and new materials for wide-ranging applications spanning, for example, electrochemistry, organic chemistry, inorganic chemistry, biochemistry, materials science and pharmaceuticals.
- Ionic liquids could contribute significantly to the development of green chemistry and green technology by, for example:

  (a) replacing toxic, flammable volatile organic solvents;
  (b) reducing or preventing chemical wastage and pollution;
  (c) improving the safety of chemical processes and products.

1.4 CATIONS AND ANIONS

Room-temperature ionic liquids are typically salts with large nitrogen- or phosphorus-containing organic cations with linear alkyl chains. Much research has been devoted to imidazolium ionic liquids, particularly those with 1-alkyl-3-methylimidazolium cations. Figure 1.1 shows the ring numbering system for these cations, along with the structures of other widely studied cations.

The following are some of the most common anions cited in the ionic liquids literature:

- halides: bromide, Br^-; chloride, Cl^-,
- nitrate, $[NO_3]^-$,
- chloroaluminates, $[AlCl_4]^-$, $[Al_2Cl_7]^-$,
- hexafluorophosphate, $[PF_6]^-$,
- tetrafluoroborate, $[BF_4]^-$,
- alkyl sulfates, $[RSO_4]^-$, for example: ethyl sulfate $[C_2H_5SO_4]^-$,

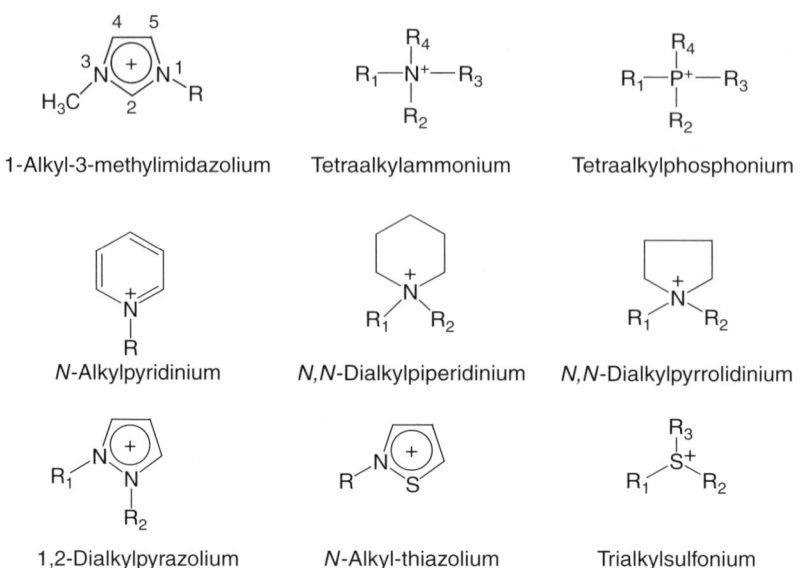

Figure 1.1 Widely studied ionic liquid cations (R, R_1, R_2, R_3 and R_4 are alkyl groups).

- alkylcarboxylates, $[RCO_2]^-$, *e.g.* acetate, $[CH_3CO_2]^-$ also written as [OAc],
- *p*-toluenesulfonate, $[CH_3C_6H_4SO_3]^-$, is also known as tosylate, $[OTs]^-$ or $[Ts]^-$,
- trifluoromethylsulfonate, $[CF_3SO_3]^-$, also known as triflate, $[OTf]^-$,
- bis(trifluoromethylsulfonyl)amide, $[N(SO_2CF_3)_2]^-$, also known as bistriflamide, or sometimes bistriflimide, $[NTf_2]^-$,
- bis(perfluoroethylsulfonyl)amide, $[N(SO_2C_2F_5)_2]^-$, shorthand: $[beti]^-$,
- dicyanamide, $[N(CN)_2]^-$ or $[dca]^-$,
- tris(pentafluoroethyl)trifluorophosphate, $[(C_2F_5)_3PF_3]^-$ or $[FAP]^-$,
- metal complexes, *e.g.* $[Co(CO)_4]^-$ and $[SbF_6]^-$.

1.5   SHORTHAND NOTATION FOR CATIONS AND ANIONS

As the names of ionic liquid cations and anions are often unwieldy, various notations and abbreviations have been employed in the scientific literature. For example, the 1-ethyl-3-methylimidazolium cation is commonly denoted as $[C_2C_1Im]^+$, $[EMI]^+$, $[EMIM]^+$, $[emim]^+$ or $[EtMeIm]^+$, with or without the brackets.

In this book 1-alkyl-3-methylimidazolium cations are denoted as $[C_nmim]^+$, where the subscript *n* is the number of carbon atoms in the linear alkyl chain. For example, the 1-ethyl-3-methylimidazolium cation is denoted as $[C_2mim]^+$.

Similarly, $[C_n py]^+$ is used as shorthand for 1-alkylpyridinium cations. For example, $[C_6 py]^+$ denotes 1-hexylpyridinium.

The C denoting the carbon atom is often dropped in the notation for tetraalkylammonium and tetraalkylphosphonium cations. For example, $[N_{1\ 8\ 8\ 8}]^+$ denotes methyltrioctylammonium and $[P_{4\ 4\ 4\ 14}]^+$ denotes tributyl(tetradecyl)phosphonium.

Square brackets are drawn around polyatomic anions but not mono-atomic anions. For example, chloride is written Cl^- whereas formate is written $[HCO_2]^-$.

When a cation is paired with an anion, the positive and negative signs of the ions are dropped in the abbreviated nomenclature. For example, 1-butyl-3-methylimidazolium chloride is written $[C_4 mim]Cl$, and 1-ethylpyridinium tetrafluoroborate is written $[C_2 py][BF_4]$.

1.6 APROTIC AND PROTIC IONIC LIQUIDS

Most research on ionic liquids has focused on aprotic ionic liquids. These are generally salts consisting solely of cations, which are not protonated, and anions. Examples are $[C_2 mim][BF_4]$ and $[C_4 mim][NTf_2]$.

Protic ionic liquids are formed by proton transfer from an acid that can donate a proton, *i.e.* a Brønsted acid (HA), to a base that can accept a proton, *i.e.* a Brønsted base (B):

$$HA + B \rightarrow [BH]^+ + A^-$$

The classic example of a protic ionic liquid is ethylammonium nitrate, $[C_2H_5NH_3][NO_3]$, which is formed by the protonation of ethylamine:

$$C_2H_5NH_2 + HNO_3 \rightarrow [C_2H_5NH_3]^+ + [NO_3]^-$$

Another example is 1-methylimidazolium tetrafluoroborate (**1.1**, $[Hmim][BF_4]$), which is prepared by the reaction of 1-methylimidazole and tetrafluoroboric acid.[1] $[Hmim][BF_4]$ is a strong Brønsted acid.

1.1

Protic ionic liquids do not necessarily consist entirely of ions.[2] Such liquids may contain a small percentage of molecular species if the proton transfer is incomplete.

1.7 BINARY MIXTURES

Many ionic liquids of interest to chemists are binary mixtures of an organic salt and an inorganic salt. They are typically binary haloaluminate systems.

Binary mixtures of the organic salt [C_2mim]Cl and aluminium chloride, $AlCl_3$, denoted [C_2mim]Cl-$AlCl_3$, are classic examples of such systems. A mixture containing [C_2mim]Cl and the Lewis acidic aluminium chloride, $AlCl_3$, in the mole ratio 1:2 is liquid at room temperature. The mixture is acidic. The cation and anion in this mixture are [C_2mim]$^+$ and the Lewis acidic [Al_2Cl_7]$^-$, respectively.

A [C_2mim]Cl-$AlCl_3$ mixture with a mole ratio of 2:1 is also a room-temperature ionic liquid. In this case, the mixture is basic. The cation is [C_2mim]$^+$, but there are two anions present: Cl^- and [$AlCl_4$]$^-$.

[C_2mim]Cl-$AlCl_3$ is an aprotic ionic liquid. Like other haloaluminate systems, it is highly sensitive to moisture.

1.8 ORGANIC AND INORGANIC IONIC LIQUIDS

Most ionic liquids reported in the literature consist of organic cations and organic or inorganic anions. For example, [C_4mim][NTf_2] consists of organic cations and organic anions, and [C_2py][BF_4] of organic cations and inorganic anions.

Inorganic ionic liquids with low melting points are also known. For example, hydrazinium bromide, [N_2H_5]Br, and hydrazinium nitrate, [N_2H_5][NO_3], melt at 86.5 and 70 °C, respectively. A binary mixture of lithium nitrate and ammonium nitrate, $LiNO_3$–NH_4NO_3, has a eutectic temperature of 98 °C.

The protic molten salt [NH_4][HF_2] almost falls within the modern definition of ionic liquids. It has a melting point of 125 °C. Another example is ammonium hydrogen sulfate, [NH_4][HSO_4], which has a melting point of 116 °C.

1.9 DEEP EUTECTIC SOLVENTS

Deep eutectic solvents, also known as eutectic-based ionic liquids, are characterized by the marked depression of freezing point when the two components of a eutectic mixture are mixed. They are typically formed by mixing a simple quaternary ammonium halide with an inorganic metal salt or an organic hydrogen-bond donor such as an amide or an alcohol. The inorganic salt or hydrogen-bond donor form a complex with the halide anion. As a result, the charge on the anion is delocalized and the freezing point of the mixture decreases.

Deep eutectic solvents were first reported by Abbott and co-workers in 2003.[3] The team showed that when choline chloride and urea, both of which are solids, are mixed together in a molar ratio of 1:2 a liquid mixture is formed that freezes at 12 °C. As pure compounds, choline chloride (2-hydroxyethyltrimethylammonium chloride, $[HOCH_2CH_2N(CH_3)_3]Cl$) and urea $[(NH_2)_2CO]$ melt at 302 and 133 °C, respectively.

Most deep eutectic solvents reported in the literature have been formed by mixing choline chloride, or other substituted quaternary ammonium salts, and metal halides. For example, eutectic mixtures of zinc chloride and substituted quaternary ammonium salts have freezing point depressions of up to 270 °C. The fluid mixtures consist of the quaternary ammonium cation and a complex of the chloride ion. The ionic mixtures dissolve metallic compounds such as nickel, copper and zinc oxides and have been exploited for the electrodeposition of metals and alloys (Section 7.6).[4]

1.10 TASK-SPECIFIC IONIC LIQUIDS

Task-specific ionic liquids are ionic liquids designed with functionalized cations and/or anions that impart specific properties or reactivities to the ionic liquid.[5] Task-specific ionic liquids are also known as functionalized ionic liquids. They are, in effect, designer ionic liquids.

Ionic liquids with imidazolium (**1.2**) or triphenylphosphine (**1.3**) cations functionalized with sulfonic acid (-SO_3H) groups are examples. These Brønsted acidic ionic liquids were first reported by Forbes, Davis, Jr., and co-workers in 2002 and used as dual solvent-catalyst systems for a range of acid-catalysed organic reactions such as esterification.[6]

H_9C_4—N⊕N—$(CH_2)_4$—SO_3H

1.2

Ph
Ph—P⁺—$(CH_2)_3$—SO_3H
Ph

1.3

1-Butyl-3-imidazolium cobalt tetracarbonyl, $[C_4mim][Co(CO_4)]$, was one of the first examples of a task-specific ionic liquid with a functionalized anion to be synthesized. Dyson, Welton and co-workers described its preparation and use as a transition-metal carbonyl catalyst for the debromination of 2-bromoketones in 2001.[7]

In 2006, Dyson and colleagues observed that numerous ionic liquids with task-specific cations have been synthesized but less effort has been devoted to the synthesis of ionic liquids with task-specific anions.[8]

1.11   CHIRAL IONIC LIQUIDS

Numerous ionic liquids have been synthesized with either chiral cations or chiral anions.[9] Chiral ionic liquids are potentially useful as solvents or chiral catalysts for asymmetric organic synthesis. Other potential applications include resolution of racemates by co-crystallization or extraction and their use as mobile or stationary phases in chromatography.

Much of the research activity on chiral ionic liquids has focused on ionic liquids with chiral cations. One example is the chiral ionic liquid di(1-phenylethyl)imidazolium nitrate, [dpeim][NO$_3$], the cation of which can exist as the (R) optical isomer or the (S)-isomer (**1.4**).[10] The cation was prepared using the chiral amine 1-phenylethylamine.

1.4

There have been relatively few reports of ionic liquids with chiral anions. The earliest example, [C$_4$mim][lactate] (**1.5**), was reported by Seddon and co-workers in 1999.[11] The ionic liquid was prepared by anion exchange between [C$_4$mim]Cl (**1.6**) and sodium (S)-2-hydroxypropionate (**1.7**) in acetone (Scheme 1.1). The sodium chloride produced by the reaction was removed by filtration and the acetone was removed by evaporation.

A rare example of an ionic liquid with a chiral cation and a chiral anion was reported by Machado and Dorta in 2005.[12] The ionic liquid was prepared by exchanging the tosylate anion of 1-methyl-3-[(S)-2'-methylbutyl]imidazolium tosylate with the chiral anion (S)-10-camphorsulfonate (Section 3.9).

1.12   GENERATIONS OF IONIC LIQUIDS

Ionic liquids are sometimes classified historically into three generations.

The haloaluminate ionic liquids comprise the first generation. The earliest examples of room-temperature ionic liquids in this category are

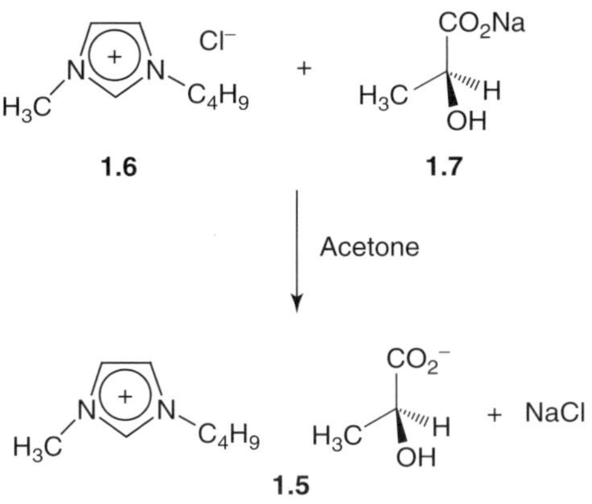

Scheme 1.1 Preparation of [C₄mim][lactate].

the eutectic mixtures of aluminium chloride and ethylpyridinium halides described by Hurley and Wier in 1951.[13] The category includes the dialkylimidazolium chloroaluminates room-temperature ionic liquids that were first reported by Wilkes, Hussey and co-workers in 1982.[14] They are formed by mixing aluminium chloride and a dialkylimidazolium chloride.

Haloaluminate ionic liquids have been studied extensively as solvents and catalysts for Friedel–Crafts and other reactions in organic chemistry (Chapters 10 and 11). These ionic liquids, however, react with water and therefore have to be handled in a dry-box.

Second-generation ionic liquids are the non-haloaluminate ionic liquids that can be used on the bench top. They have been used extensively as solvents in organic chemistry. The first examples were described by Wilkes and Zaworotko in 1992.[15] They include dialkylimizadolium ionic liquids with weakly coordinating anions such as hexafluorophosphate $[PF_6]^-$ and tetrafluoroborate $[BF_4]^-$. These ionic liquids were originally thought to be stable in air and water but it was subsequently shown that they undergo hydrolysis under certain conditions, resulting in the formation of toxic and corrosive hydrogen fluoride.[16]

In 1996, Bonhôte, Grätzel and co-workers described the synthesis of hydrophobic dialkylimidazolium ionic liquids with $[NTf_2]^-$ anions.[17] These second-generation ionic liquids are stable in the presence of air and water.

The third generation are the task-specific ionic liquids and chiral ionic liquids described above that emerged on the research scene in the early 2000s.

The three generations are broad categories and not exclusive. The generations generally refer to pyridinium and particularly imidazolium ionic liquids as these are the families of ionic liquids that have been studied most extensively in recent years.

Other types of ionic liquids do not fall conveniently into these categories. For example, ethylammonium nitrate is not considered to be a first-generation ionic liquid even though it was first reported in 1914, *i.e.* before the emergence of first-generation ionic liquids.[18]

REFERENCES

1. H. P. Zhu, F. Yang, J. Tang and M.-Y. He, *Green Chem.*, 2003, **5**, 38.
2. T. L. Greaves and C. J. Drummond, *Chem. Rev.*, 2008, **108**, 206.
3. A. P. Abbott, G. Capper, D. L. Davies, R. K. Rasheed and V. Tambyrajah, *Chem. Commun.*, 2003, **70**.
4. A. P. Abbott, G. Capper, K. J. McKenzie and K. S. Ryder, *J. Electroanal. Chem.*, 2007, **599**, 288.
5. J. H. Davis Jr., *Chem. Lett.*, 2004, 1072.
6. A. C. Cole, J. L. Jensen, I. Ntai, K. L. T. Tran, K. J. Weaver, D. C. Forbes and J. H. Davis Jr., *J. Am. Chem. Soc.*, 2002, **124**, 5962.
7. R. J. C. Brown, P. J. Dyson, D. J. Ellis and T. Welton, *Chem. Commun.*, 2001, 1862.
8. Z. Fei, T. J. Geldbach, D. Zhao and P. J. Dyson, *Chem. Eur. J.*, 2006, **12**, 2122.
9. C. Baudequin, D. Brégon, J. Levillain, F. Guillen, J.-C. Plaquevent and A.-C. Gaumont, *Tetrahedron: Asymmetry*, 2005, **16**, 3921.
10. A. J. Carmichael, M. Deetlefs, M. J. Earle, U. Fröhlich and K. R. Seddon, in *Ionic Liquids as Green Solvents Progress and Prospects*, ed. R. D. Rogers and K. R. Seddon, ACS Symposium Series 856, American Chemical Society, Washington DC, 2003, p. 14.
11. M. J. Earle, P. B. McCormac and K. R. Seddon, *Green Chem.*, 1999, **1**, 23.
12. M. Y. Machado and R. Dorta, *Synthesis*, 2005, 2473.
13. F. H. Hurley and T. P. Wier, *J. Electrochem. Soc.*, 1951, **98**, 203 and 207.
14. J. S. Wilkes, J. A. Levisky, R. A. Wilson and C. L. Hussey, *Inorg. Chem.*, 1982, **21**, 1263.
15. J. S. Wilkes and M. J. Zaworotko, *J. Chem. Soc., Chem. Commun.*, 1992, 965.

16. R. P. Swatloski, J. D. Holbrey and R. D. Rogers, *Green Chem.*, 2003, **5**, 361.
17. P. Bonhôte, A.-P. Dias, N. Papageorgiou, K. Kalyanasundaram and M. Grätzel, *Inorg. Chem.*, 1996, **35**, 1168.
18. P. Walden, *Bull. Acad. Imper. Sci (St Petersburg)*, 1914, 1800.

CHAPTER 2
History

2.1 ORIGINS

The origins of modern ionic liquids chemistry, that is the chemistry of salts with low melting points, can be traced back to the second half of the nineteenth century when chemists noted that a liquid, known as "red oil", often appeared as a separate phase during Friedel–Crafts reactions.

Friedel and Crafts first described their alkylation and acylation reactions in a French scientific journal in 1877.[1] They noted that when small quantities of anhydrous aluminium chloride were added to amyl chloride a reaction ensued that resulted in a liquid that divided into two layers.

It was not until almost a century later that chemists in Japan showed that the red oil consisted of an alkylated aromatic ring cation and a chloroaluminate anion.[2] The oil was, therefore, an ionic liquid.

In 1888, Gabriel reported the discovery of the protic ionic liquid ethanolammonium nitrate.[3] The liquid has a melting point of 52–55 °C and was probably the first organic salt with a melting point of less than 100 °C to be reported, according to the authors of a review of protic ionic liquids published in 2008.[4]

By the end of the nineteenth century, there was some interest in the synthesis and study of other salts with low melting points. For example, Trowbridge and co-workers published a series of papers on the halides and perhalides of pyridine.[5-7] In one of these papers, the chemists reported that the melting point of "pyridine hydrochloride, $C_5H_5N.HCl$" was 82 °C, which would make the compound an ionic liquid by the modern definition. They prepared the compound by adding concentrated

An Introduction to Ionic Liquids
By Michael Freemantle
© Michael Freemantle 2010
Published by the Royal Society of Chemistry, www.rsc.org

hydrochloric acid in excess to pure pyridine. "The salt separates out as a mass of thin scales, which are purified by recrystallization from absolute alcohol", the authors observed.

Almost 40 years later, the compound went by the name "pyridinium hydrochloride" and its melting point was disputed. It melted at 144.5 °C "as contrasted with the melting point found by Trowbridge and Diehl" of 82 °C, noted the authors of a paper entitled: "Fused 'Onium' Salts as Acids. Reactions in Fused Pyridinium Hydrochloride".[8] Onium compounds contain cations derived by the addition of H^+ to a parent hydride with one atom of an element such as nitrogen, oxygen or phosphorus. For example, ammonium chloride contains the cation $[NH_4]^+$, which is derived by adding H^+ to ammonia, NH_3.

The authors concluded, in their 1936 paper, that purified pyridinium hydrochloride is a typical "onium" salt that possesses acidic character in the fused state, reacting "with metals and metallic oxides much as does hydrochloric acid in aqueous solution". They added that metallic chlorides are soluble in fused pydrinuium chloride and form a stable low temperature melt. "Such melts conduct the electric current with facility and in the case of tin, lead, arsenic, antimony, bismuth, and mercury permit the electrodeposition of these metals", they explained.

Over the next three decades, several studies confirmed that pyridinium hydrochloride has a melting point of 144.5 °C, making it a low-temperature rather than room-temperature fused salt. Goffman and Harrington observed that it is "virtually impossible" to maintain the salt as a melt for long periods of time without "appreciable" decomposition.[9] Even so, fused pyridinium chloride has been used as a solvent, *e.g.* to investigate complexes of Co(II) spectroscopically, the two authors pointed out.

Nowadays, it is generally accepted that the birth of room-temperature ionic liquids took place in 1914 with Walden's synthesis of ethyl-ammonium nitrate. Progress was subsequently slow, with the next significant development not being reported until 1951. The explosion of interest in ionic liquids commenced in the years 1998–2000, resulting in an exponential increase in research publications on the topic during the first decade of the new millennium. Some of the most significant developments in the chemistry of room-temperature ionic liquids since 1998 are outlined in the following chapters of this book.

2.2 KEY DATES IN THE HISTORY OF IONIC LIQUIDS

1888. Gabriel reports the discovery of the protic ionic liquid ethanol-ammonium nitrate, which has a melting point of 52–55 °C.[3] It is possibly the first example of an organic salt with a melting point of less than 100 °C.

1914. Walden publishes the synthesis of the protic ionic liquid ethylammonium nitrate, which has a melting point of 12.5 °C.[10] It is probably the first room-temperature ionic liquid to be described in the scientific literature.

1934. Graenacher uses molten 1-ethylpyridinium chloride, in the presence of nitrogen-containing bases, to help dissolve cellulose.[11] The system has a melting point of 118 °C.

1951. Hurley and Wier report the discovery, patented in 1948, that mixtures of aluminium chloride and ethylpyridinium halides are molten at room temperature.[12] They used the melts as electrolytes for the electrodeposition of aluminium.

1961. Bloom coins the term "ionic liquids" in a lecture at a Faraday Society Discussion on "The Structure and Properties of Ionic Melts".[13] He used the term to refer to pure molten salts that consist predominantly of ions. The lecture focused exclusively on molten alkali chlorides such as LiCl.

1963. Yoke III and co-workers report the synthesis of triethyl-ammonium dichlorocuprate(I), $[(C_2H_5)_3NH][CuCl_2]$, which is a light green oil at 25 °C and therefore a room-temperature ionic liquid.[14]

1964. Gordon suggests that melts of low-melting point quaternary ammonium salts could be interesting media for organic reactions.[15]

1970. Bockris devotes a whole chapter to ionic liquids in a book on electrochemistry.[16] He notes that most fused salts are stable as liquids only at relatively high temperatures although tetraalkylammonium salts are liquid at much lower temperatures.

1972. Parshall reports the use of low-temperature molten salt media, $[N_{2\ 2\ 2\ 2}][GeCl_3]$ and $[N_{2\ 2\ 2\ 2}][SnCl_3]$, for homogeneous catalysis.[17]

1976. Osteryoung and co-workers publish a paper on Friedel–Crafts alkylations in a room-temperature molten-salt medium: a melt of aluminium chloride and ethylpyridinium bromide.[18]

1978. The Osteryoung group describes an aluminium chloride/1-butylpyridinium chloride salt, $[C_4py]Cl-AlCl_3$, system that is molten at ambient temperatures over a wide range of compositions.[19]

1981. Knifton reports the use of phosphonium ionic liquids to synthesize ethylene glycol from synthesis gas $(CO + H_2)$.[20]

1982. Wilkes, Hussey and others report the discovery of a "new class of room-temperature ionic liquids", the dialkylimidazolium chloroaluminates.[21]

Poole and co-workers show that ethylammonium nitrate is suitable for use as a stationary phase in gas–liquid chromatography.[22]

Seddon submits a proposal to the Science & Engineering Research Council in the UK to fund a project to investigate the use of room-temperature ionic liquids for catalytic processes that might have some industrial application. The proposal receives a gamma (low) rating and is rejected. One referee suggests the chemistry is too complicated for potential applications. Another suggests the chemistry is so trivial it is not worth doing.[23]

1984. Magnuson and co-workers publish the first study of ionic liquid–enzyme systems.[24]

1989. The BP Venture Research Unit awards Seddon's group at the University of Sussex, UK, a substantial grant to investigate the chemistry of ionic liquids (D. Braben, personal communication, 1 November 2007).

1990. Chauvin and co-workers describe the catalytic dimerization of alkenes by nickel complexes in dialkylimidazolium, alkylpyridinium and tetraalkylphosphonium chloroaluminate molten salts.[25]

1992. Wilkes and Zaworotko report the preparation of "air and water stable" 1-ethyl-3-methylimidazolium ionic liquids with weakly coordinating anions such as hexafluorophosphate $[PF_6]^-$ and tetrafluoroborate $[BF_4]^-$.[26]

1995. Chauvin and co-workers show that rhodium complexes dissolved in dialkylimidazolium ionic liquids with non-nucleophilic anions catalyse the hydrogenation, isomerization and hydroformylation of pentene.[27]

1996. Eastman Chemical launches a process for the synthesis of 2,5-dihydrofuran using a tetraalkylphosphonium ionic liquid (Section 14.2).

Bonhôte, Grätzel and co-workers report the synthesis of hydrophobic dialkylimidazolium ionic liquids with $[NTf_2]^-$ anions.[28]

1997. Seddon publishes a paper on the use of ionic liquids for clean technology.[29]

Howarth describes the first example of a chiral ionic liquid.[30]

1998. The French Petroleum Institute (IFP) announces the launch of the commercial Difasol process for the production of iso-octenes from *n*-butene using a nickel catalyst dissolved in a chloroaluminate ionic liquid.[23]

Chemical & Engineering News publishes an article entitled "Designer Solvents: Ionic liquids may boost clean technology development"[23] that sparks substantial interest in room-temperature ionic liquids.[31]

1999. The Queen's University Ionic Liquids Laboratories (QUILL) Research Centre is opened in Belfast, Northern Ireland.[32]

2000. A NATO (North Atlantic Treaty Organization) advanced research workshop on "Green Industrial Applications of Ionic Liquids" is held in Crete.[33] The meeting attracts nearly every active researcher in ionic liquids in the world: a mere 65 researchers.

Davis, Jr. introduces the "task-specific ionic liquid" concept at the NATO meeting in Crete and at other venues.[34]

2002. Rogers and co-workers show that certain imidazolium ionic liquids dissolve cellulose pulp.[35]

2003. [C₄mim][PF₆], one of the most widely used ionic liquids, is shown not to be "green".[36] The anion undergoes hydrolysis leading to the evolution of acidic HF fumes.

BASF announces an industrial process that uses ionic liquids to scavenge acids from reaction mixtures (Section 14.4).

2005. The 1st International Congress on Ionic Liquids, held in Salzburg, Austria, attracts more than 400 participants from 33 countries.[37]

Rogers, Seddon and Holbrey are among the recipients of the 2005 Presidential Green Chemistry Challenge Awards administered by the Environmental Protection Agency in the USA. The three chemists received their awards for developing ionic liquids as recyclable solvents to dissolve and process cellulose into useful materials.

2006. QUILL is one of 21 winners of the Queen's Anniversary Prizes for Higher & Further Education in the UK.

2007. The number of papers and patents on ionic liquids passes the 8000 mark, 97% of which have been published since 1998.[31]

2008. Interest in ionic liquids continues to grow. Over 2500 papers on the topic are published during the year, compared with just 37 in 1999 and 76 in 2000 (M. Fanselow, personal communication, 28 January 2009).

REFERENCES

1. C. Friedel and J. M. Crafts, *Compt. Rend.*, 1877, **84**, 1392 and 1450.
2. N. Nambu, N. Hiraoka, K. Shigemura, S. Hamanaka and M. Ogawa, *Bull. Chem. Soc. Jpn.*, 1976, **49**, 3637.
3. S. Gabriel, *Ber.*, 1888, **21**, 2669.
4. T. L. Greaves and C. J. Drummond, *Chem. Rev.*, 2008, **108**, 206.
5. A. B. Prescott and P. F. Trowbridge, *J. Am. Chem. Soc.*, 1895, **17**, 859.
6. P. F. Trowbridge, *J. Am. Chem. Soc.*, 1897, **19**, 322.
7. P. F. Trowbridge and O. C. Diehl, *J. Am. Chem. Soc.*, 1897, **19**, 558.

8. L. F. Audrieth, A. Long and R. E. Edwards, *J. Am. Chem. Soc.*, 1936, **58**, 428.
9. M. Goffman and G. W. Harrington, *J. Phys. Chem.*, 1963, **67**, 1877.
10. P. Walden, *Bull. Acad. Imper. Sci. (St Petersburg)*, 1914, 1800.
11. C. Graenacher, *U. S. Pat.* 1,943,176 (Jan. 9, 1934).
12. F. H. Hurley and T. P. Wier, *J. Electrochem. Soc.*, 1951, **98**, 203 and 207.
13. H. Bloom, *Discuss. Faraday Soc.*, 1961, **32**, 7.
14. J. T. Yoke III, J. F. Weiss and G. Tollin, *Inorg. Chem.*, 1963, **2**, 1210.
15. J. E. Gordon, *J. Am. Chem. Soc.*, 1964, **86**, 4492.
16. J. O'M. Bockris and A. K. N. Reddy, *Modern Electrochemistry*, Plenum, New York, 1970, p. 513.
17. G. W. Parshall, *J. Am. Chem. Soc.*, 1972, **94**, 8716.
18. V. R. Koch, L. L. Miller and R. A. Osteryoung, *J. Am. Chem. Soc.*, 1976, **98**, 5277.
19. R. J. Gale, B. Gilbert and R. A. Osteryoung, *Inorg. Chem.*, 1978, **17**, 2728.
20. J. F. Knifton, *J. Am. Chem. Soc.*, 1981, **103**, 3959.
21. J. S. Wilkes, J. A. Levisky, R. A. Wilson and C. L. Hussey, *Inorg. Chem.*, 1982, **21**, 1263.
22. F. Pacholec, H. T. Butler and C. F. Poole, *Anal. Chem.*, 1982, **54**, 1938.
23. M. Freemantle, *Chem. Eng. News*, March 30, 1998, 32.
24. D. K. Magnuson, J. W. Bodley and D. F. Evans, *J. Solution Chem.*, 1984, **13**, 583.
25. Y. Chauvin, B. Gilbert and I. Guibard, *J. Chem. Soc., Chem. Commun.*, 1990, 1715.
26. J. S. Wilkes and M. J. Zaworotko, *J. Chem. Soc., Chem. Commun.*, 1992, 965.
27. Y. Chauvin, L. Mussmann and H. Olivier, *Angew. Chem. Int. Ed. Engl.*, 1995, **34**, 2698.
28. P. Bonhôte, A.-P. Dias, N. Papageorgiou, K. Kalyanasundaram and M. Grätzel, *Inorg. Chem.*, 1996, **35**, 1168.
29. K. R. Seddon, *J. Chem. Technol. Biotechnol.*, 1997, **68**, 351.
30. J. Howarth, K. Hanlon, D. Fayne and P. McCormac, *Tetrahedron Lett.*, 1997, **38**, 3097.
31. R. D. Rogers and G. A. Voth, *Acc. Chem. Res.*, 2007, **40**, 1077.
32. M. Freemantle, *Chem. Eng. News*, Jan. 4, 1999, 23.
33. M. Freemantle, *Chem. Eng. News*, May 15, 2000, 37.

34. J. H. Davis Jr., *Chem. Lett.*, 2004, 1072.
35. R. P. Swatlowski, S. K. Spear, J. D. Holbrey and R. D. Rogers, *J. Am. Chem. Soc.*, 2002, **124**, 4974.
36. R. P. Swatloski, J. D. Holbrey and R. D. Rogers, *Green Chem.*, 2003, **5**, 361.
37. M. Freemantle, *Chem. Eng. News*, Aug 1, 2005, 33.

CHAPTER 3
Synthesis of Ionic Liquids

3.1 INTRODUCTION

Since the early 2000s, numerous room-temperature ionic liquids have become commercially available in gram to multi-ton quantities. In 2003, for example, Davis, Jr. and Fox compiled a list of some 110 commercially-availably low-melting salts of potential use as ionic liquids.[1] All the listed salts have melting points of 150 °C or less and many are liquid at room temperature. Four years later, Schubert observed that "some 500 ionic liquids are produced commercially on a lab scale for R&D purposes".[2] He suggested that the number is likely to rise to tens of thousands as research interest and applications expand.

Until the late 1990s, room-temperature ionic liquids "were considered to be rare", according to Welton.[3] In those days, researchers generally had to synthesize the liquids they used themselves. Interestingly, for example, the synthesis of non-chloroaluminate ionic liquids with dialkyl-imidazolium cations was first reported in 1992.[4] [C$_2$mim][PF$_6$] did not feature in the scientific literature until 1994.[5] It would take another five years before these now widely-studied non-chloroaluminate dialkyl-imidazolium ionic liquids became commercially available.

As the number of ionic liquids increases the variety of methods for synthesizing them is also increasing. Some of the most common methods of preparing ionic liquids are outlined below.

An Introduction to Ionic Liquids
By Michael Freemantle
© Michael Freemantle 2010
Published by the Royal Society of Chemistry, www.rsc.org

3.2 ALKYLATION

The alkyl cations of many ammonium, imidazolium, pyridinium and phosphonium ionic liquids are prepared by alkylation of a suitable precursor, a nucleophile, using an alkylating agent such a halogeno-alkane or a dialkyl sulfate.

For example, [C_2mim]Cl (**3.1**) is prepared by the reaction of 1-methyl-imidazole (**3.2**) and compressed gaseous chloroethane (Scheme 3.1). The reaction can be carried out in an autoclave without the addition of a solvent. [C_2mim]Cl has a melting point of 87 °C and is therefore solid at room temperature. As its melting point is below 100 °C, it is, by defini-tion, an ionic liquid. It is also a precursor for the synthesis of other imidazolium ionic liquids.

Similarly, [C_4mim]Cl is commonly used as a precursor for preparing other ionic liquids. It is obtained by the reaction of 1-methylimidazole and 1-chlorobutane. The salt, which has a melting point of 69 °C, is recrystallized using a solvent such as ethyl ethanoate.

The 1-alkylimidazoles that are used as starting materials for the synthesis of these and other imidazolium ionic liquids are readily pre-pared by alkylation of imidazole.

Halogenalkanes are also used to alkylate pyridine in the preparation of alkylpyridinium salts. For example, [C_2py]Br is prepared by the re-action of pyridine and bromoethane.

Asymmetrical tetraalkylphosphonium halides are typically prepared by the nucleophilic addition of a tertiary phosphine to a haloalkane. For example, the room temperature ionic liquid trihexyl(tetradecyl)phosphonium chloride, [$P_{6\ 6\ 6\ 14}$]Cl, is prepared by the reaction of tri-hexylphosphine, $P(C_6H_{13})_3$, and 1-chlorotetradecane, $C_{14}H_{29}Cl$. The tertiary phosphines used for such reactions are synthesized by the addi-tion of phosphine gas, PH_3, to an α-olefin such as $CH_2=CH(CH_2)_3CH_3$.

3.3 ANION EXCHANGE

Many ionic liquids, for example the commonly-studied tetrafluoroborate and hexafluorophosphate ionic liquids with dialkylimidazolium cations,

Scheme 3.1 Synthesis of [C_2mim]Cl.

are synthesized in a two-step process. First the halide salt with the required cation is prepared by alkylation (Section 3.2). The halide anion is then exchanged with the required anion, typically by anion metathesis.

In anion metathesis preparations, anions are exchanged between an organic salt and an inorganic anion source such as a group 1 metal salt or a silver salt.

Ionic liquids with the $[NTf_2]^-$ anion, for example, are prepared by the metathetic reaction between an organic halide salt and lithium bis[(trifluoromethyl)sulfonyl]amide. For example, the dialkylimidazolium ionic liquid $[C_4mim][NTf_2]$, the alkylpyridinium ionic liquid $[C_2py][NTf_2]$ and the tetraalkylphosphonium ionic liquid $[P_{6\ 6\ 6\ 14}][NTf_2]$ are prepared as follows:

$$[C_4mim]Cl + Li[NTf_2] \rightarrow [C_4mim][NTf_2] + LiCl$$

$$[C_2py]Cl + Li[NTf_2] \rightarrow [C_2py][NTf_2] + LiCl$$

$$[P_{6\,6\,6\,14}]Cl + Li[NTf_2] \rightarrow [P_{6\,6\,6\,14}][NTf_2] + LiCl$$

Similarly, ionic liquids with the triflate anion can be prepared using sodium triflate, for example:

$$[C_4mim]Cl + Na[OTf] \rightarrow [C_4mim][OTf] + NaCl$$

Silver nitrate and other silver salts are widely used to prepare a range of ionic liquids. For example, $[C_2mim][NO_3]$ is prepared by the reaction of $[C_2mim]I$ and silver nitrate:

$$[C_2mim]I + Ag[NO_3] \rightarrow [C_2mim][NO_3] + AgI$$

In 2005, Drake and co-workers reported the use of silver nitrate to prepare 1-alkyl-4-amino-1,2,4-triazolium nitrate ionic liquids such as **3.3** (Scheme 3.2).[6] The metathesis reactions were carried out in hot methanol and the silver halides produced by the reactions were removed by filtration. Many of the ionic liquid products, notably those with shorter alkyl chains, were liquid at room temperature. The 1-alkyl-4-amino-1,2,4-triazolium halide starting material **3.4** was prepared by the alkylation of 4-amino-1,2,4-triazole (**3.5**) with excess halogenoalkane, using a solvent such as acetonitrile.

In 2000, Reed and co-workers described the synthesis of dialkylimidazolium ionic liquids with carborane anions, such as $[C_2mim][CB_{11}H_{12}]$,

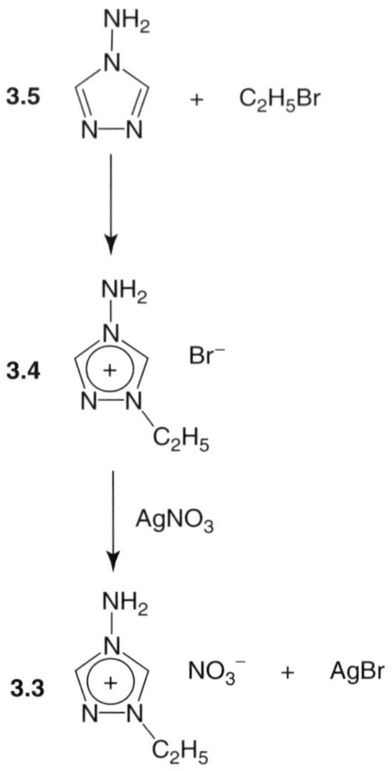

Scheme 3.2   Alkylation and anion metathesis.

by anion metathesis using silver salts:[7]

$$[C_2mim]Cl + Ag[CB_{11}H_{12}] \rightarrow [C_2mim][CB_{11}H_{12}] + AgCl$$

The dialkylimidazolium carborane ionic liquids prepared by the group have melting points in the range 45–156 °C. Carborane anions are of interest in the study of ionic liquids because they are among the most inert and least nucleophilic anions known.

Dialkylimidazolium ionic liquids with $[PF_6]^-$ anions are prepared by reaction of a dialkylimidazolium halide with an aqueous solution of the acid HPF_6 or a neutral salt such as $Na[PF_6]$ or $[NH_4][PF_6]$, for example:

$$[C_4mim]Cl + HPF_6 \rightarrow [C_4mim][PF_6] + HCl$$

As $[C_4mim][PF_6]$ is immiscible with water, it forms a separate phase that can be readily separated from the aqueous layer.

Ion-exchange resins can also be used to prepared ionic liquids. For example, the room-temperature ionic liquid 1-ethyl-3-methylimidazolium vinylsulfonate, $[C_2mim][CH_2{=}CH{-}SO_3]$, is prepared by flushing an aqueous solution of [C$_2$mim]Br through a commercial ion exchange resin loaded with [CH$_2$=CH-SO$_3$]$^-$ anions.[8]

3.4 MICROWAVE- AND ULTRASOUND-ASSISTED SYNTHESIS

In 2001, Varma and Namboordiri noted that the synthesis of dialkyl-imidazolium halides typically involves refluxing 1-methylimidazole and a large excess of the alkyl halide in an organic solvent such as toluene for several hours.[9] The two researchers showed that this tedious procedure could be side-stepped by using microwave-assisted synthesis. They prepared a range of dialkylimidazolium halides in an unmodified household microwave oven with reaction times of just a few minutes. The procedure avoids the use of solvents and a large excess of the alkyl halides.

Varma subsequently showed that dialkylimidazolium ionic liquids with tetrafluoroborate anions, for example [C$_4$mim][BF$_4$], can also be synthesized in a microwave oven under solvent-free conditions:[10]

$$[C_4mim]Cl + [NH_4][BF_4] \rightarrow [C_4mim][BF_4] + NH_4Cl$$

In 2002, Lévêque and co-workers described the use of an ultrasonic reactor for the preparation of [C$_4$mim][BF$_4$], [C$_4$mim][PF$_6$] and related ionic liquids.[11] The sonochemical synthesis resulted in yields of up to 90%, a "dramatic" reduction of the reaction time and improved quality of the products, according to the authors.

3.5 HALIDE-FREE SYNTHESIS

As we saw above, the synthesis of non-haloaluminate ionic liquids typically involves two steps, the first of which uses a halogenoalkane as an alkylating agent. The second step, anion metathesis, generates un-wanted halide salts.

Halogenoalkanes, particularly those with high boiling points, are difficult to separate from the products at the end of a reaction. In addition, halide salts produced by the metathesis reactions are likely to contaminate the ionic liquids, even after they have been precipitated from the reaction mixtures and removed by filtration. Such contamin-ation can have a dramatic impact on the physical properties of the ionic liquid (Section 4.15). Halide salt impurities may also poison and

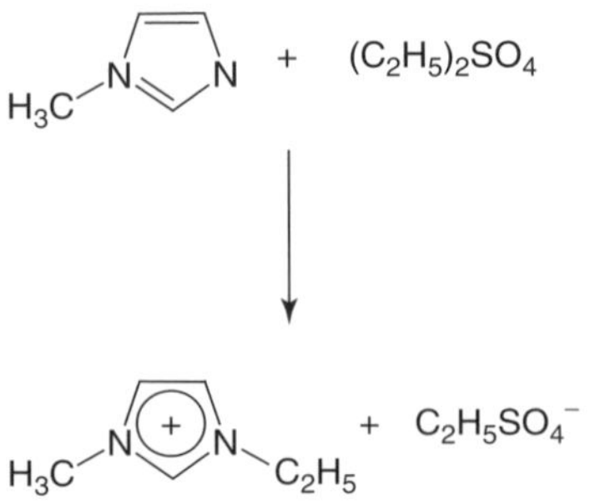

3.6

Scheme 3.3   Halide-free synthesis.

deactivate catalysts when ionic liquids are used as solvents for transition metal catalysis.

Consequently, the possibility of developing halide-free syntheses of ionic liquids has been explored by several research groups since the early 2000s. In one such study, Holbrey and co-workers prepared several 1,3-dialkylimidazolium alkyl sulfate ionic liquids, *e.g.* **3.6**, by the alkylation of alkylimidazoles with dimethyl sulfate or diethyl sulfate, $(C_2H_5)_2SO_4$ (Scheme 3.3).[12] These two alkylating agents are commonly used in industry.

Most alkyl sulfate ionic liquids are liquid at room temperature and stable at up to 370 °C or higher. The ionic liquids can be used to prepare other ionic liquids by metathesis. For example, Holbrey's team have prepared [C_4mim][PF_6] by the metathesis reaction of the methane-sulfonate ionic liquid [C_4mim][CH_3SO_4] with hexafluorophosphoric acid, HPF_6.

Dimethyl carbonate, $(CH_3O)_2CO$, has also been used as an alkylating agent to prepare halogen-free ionic liquids.[13] The methylation of 1-ethyl-2-methyl-imidazoline (**3.7**), for instance, yields an ionic liquid (**3.8**) with the 1-ethyl-2,3-dimethylimidazolinium cation and the methyl monocarbonate anion (Scheme 3.4).

3.6   SYNTHESIS BY PROTONATION

Protic ionic liquids (Section 1.6) are prepared by protonation of a starting material such as an amine or phosphine. For example, ethyl-ammonium nitrate is prepared by adding concentrated nitric acid to an aqueous solution of ethylamine, $C_2H_5NH_2$. The water is removed by distillation. $[C_2H_5NH_3][NO_3]$, like other protic ionic liquids, is subject to decomposition by deprotonation.

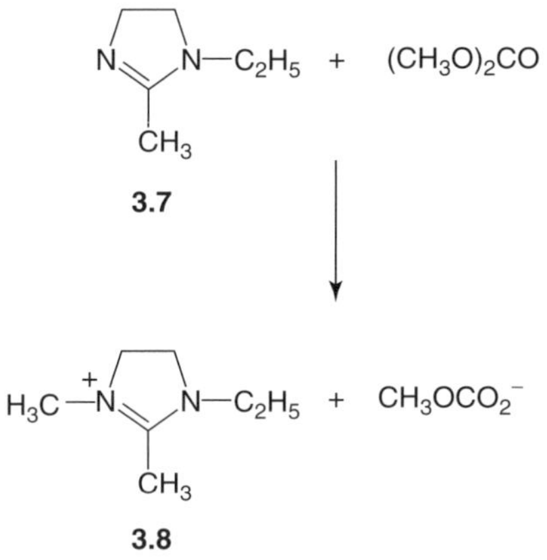

3.7

3.8

Scheme 3.4 Alkylation with dimethyl carbonate.

3.7 SYNTHESIS OF HALOALUMINATE IONIC LIQUIDS

Haloaluminate ionic liquids are prepared by the reaction of an organic halide salt with the required cation and a Lewis acid. Typically, a 1-alkyl-3-methylimidazolium chloride, $[C_n\text{mim}]Cl$, is mixed with Lewis acidic aluminium chloride:

$$[C_n\text{mim}]Cl + AlCl_3 \rightarrow [C_n\text{mim}][AlCl_4] \quad ([C_n\text{mim}]Cl \text{ in excess})$$

$$[C_n\text{mim}]Cl + AlCl_3 \rightarrow [C_n\text{mim}][Al_2Cl_7] \quad (AlCl_3 \text{ in excess}).$$

3.8 SYNTHESIS OF TASK-SPECIFIC IONIC LIQUIDS

Ionic liquids with functionalized cations and/or functionalized anions have a wide range of potential applications.[14] Functionalized ionic liquids, or task-specific ionic liquids as they are also known, are potentially useful as reaction media for chemical synthesis, as catalysts and as materials as diverse as lubricants, rocket propellants and active pharmaceutical ingredients (Chapter 14). Their synthesis is therefore a rapidly growing area of research.

Various synthetic routes have been reported for the preparation of task-specific ionic liquids. Typically, functionalized halogenoalkanes are used to prepare ionic liquids with functionalized imidazolium, phosphonium and pyridinium cations. One of the earliest reports focused on the development of task-specific imidazolium ionic liquids for the extraction of metal ions from aqueous solutions (Section 5.17).[15] The authors of the report used 1-(3-aminopropyl)imidazole (**3.9**) as a starting material and a two-step procedure to synthesize six ionic liquids with cations functionalized with urea, thiourea and thioether groups. In the first step the amine group of the starting material was converted into the functional group. In the second step, the nitrogen atom at the 3-position in the imidazole ring was alkylated using an alkyl halide.

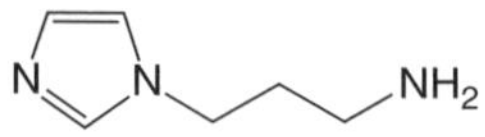

3.9

Various ionic liquids with functionalized anions have been prepared by anion exchange with commercially available alkali salts and acids. For example, in 2005 Ohno and co-workers described the preparation of room-temperature ionic liquids from 20 natural amino acids.[16] The team first prepared an aqueous solution of [C_2mim][OH] (**3.10**) from [C_2mim]Br using an anion exchange resin. The solution was then added to an aqueous solution containing a slight excess of the amino acid, *e.g.* alanine (**3.11**, Scheme 3.5). The water was then removed by evaporation and the excess amino acid by filtration.

Other methods for functionalizing anions include transesterification. The method has been used to prepare ionic liquids with functionalized alkyl sulfate anions. Wasserscheid and co-workers used transesterification with ethylene glycol-based alcohols, such as 2-methoxyethanol, $CH_3O(CH_2)_2OH$, under slightly acidic conditions to replace the alkyl

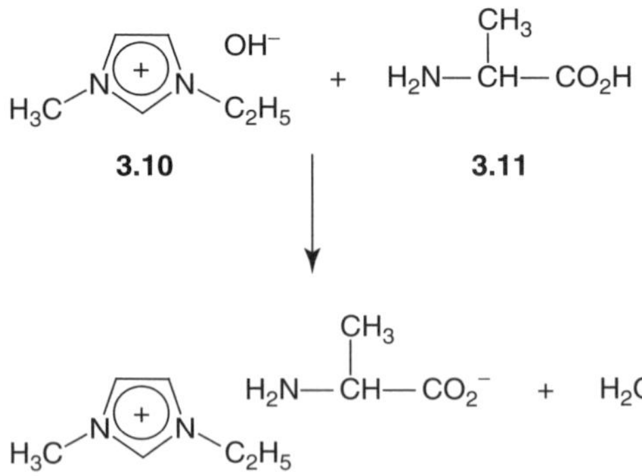

Scheme 3.5 Synthesis of an amino acid ionic liquid.

groups in the anions of [C₂mim][CH₃OSO₃] and [C₂mim][C₂H₅OSO₃] with ethylene-glycol functionalities,[17] for example:

$$[C_2mim][C_2H_5OSO_3] + CH_3O(CH_2)_2OH$$
$$\rightarrow [C_2mim][CH_3O(CH_2)_2OSO_3] + C_2H_5OH$$

The ethanol was continuously removed by evaporation during the process. The ionic liquid, which does not evaporate because it has negligible vapour pressure, was produced in around 99% yield. It has a melting point of less than 0 °C.

3.9 CHIRAL SYNTHESIS

Chiral ionic liquids are potentially important as chiral solvents and catalysts in asymmetric synthesis. They are derived from the chiral pool. This pool consists of a wide range of chiral precursors that include synthetic enantiomers that are produced on a large scale industrially and chiral amino acids, carbohydrates, terpenes and other chiral compounds that occur naturally in animals and plants.

Following the work by Ohno's group in 2005 (Section 3.8),[16] amino acids have been exploited as a source of chiral anions for ionic liquids in various investigations. However, many of the syntheses of chiral ionic liquids have focused on ionic liquids with chiral cations. These ionic liquids are typically prepared by alkylation of a chiral precursor followed by anion exchange.

In 2004, for example, Mauduit, Guillemin and co-workers reported the synthesis of a room-temperature ionic liquid consisting of chiral cations derived from the amino acid L-valine and [NTf₂]⁻ anions.[18] The group used commercially available *N*-Boc-L-valine (**3.12**) as a starting material. Boc (*tert*-butyloxycarbonyl) is widely used as a protecting group for amine groups in peptide synthesis.

In a sequence of steps, the group coupled the protected amino acid and the substituted aniline **3.13** to form the enantiopure imidazoline **3.14**. Alkylation of this compound with alkyl halides followed by anion exchange yielded a variety of chiral imidazolinium salts, some of which are solid at room temperature, but one of which, the [NTf₂]⁻ salt (**3.15**), is liquid (Scheme 3.6).

Ionic liquids with both chiral cations and chiral anions have also been synthesized. In 2005, Machado and Dorta reported the synthesis of several chiral imidazolium salts, including the "first example of a 'doubly chiral' ionic liquid".[19] The two chemists first prepared several

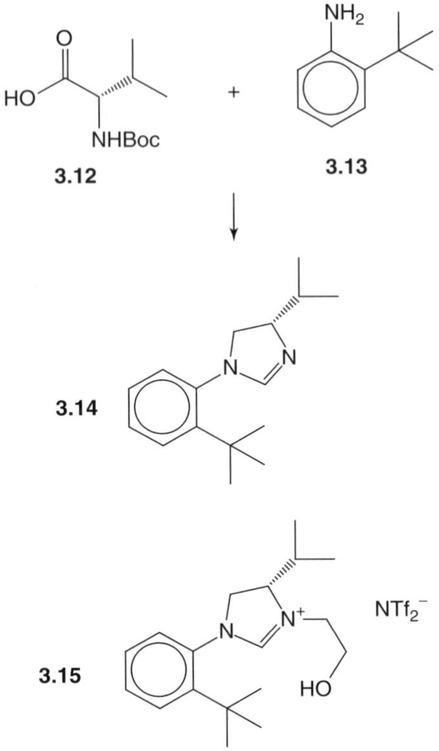

Scheme 3.6   Synthesis of an ionic liquid with a chiral imidazolinium cation.

chiral imidazolium salts derived from the chiral pool precursors cam-
phor, β-pinene and tartaric acid. They included ionic liquids with chiral
imidazolium cations and others with chiral anions such as (*S*)-10-cam-
phorsulfonate. The two chemists prepared the doubly chiral ionic liquid
3.16 by passing an aqueous solution of an ionic liquid with chiral imidazolium cations and tosylate anions over an ion-exchange resin
charged with (*S*)-camphorsulfonate anions. Stripping water from the
resulting aqueous solution and drying under vacuum yielded a pale-
yellow viscous oil.

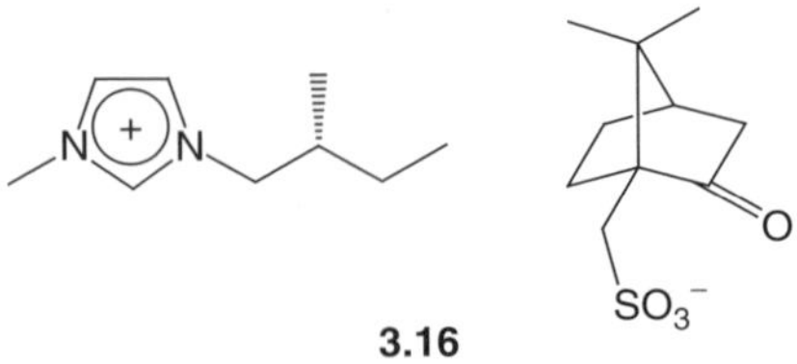

3.16

REFERENCES

1. J. H. Davis Jr. and P. A. Fox, in *Ionic Liquids as Green Solvents: Progress and Prospects*, ed. R. D. Rogers and K. R. Seddon, ACS Symposium Series 856, American Chemical Society, Washington DC, 2003, p. 100.
2. M. Freemantle, *Chem. Eng. News*, Jan. 1, 2007, 23.
3. T. Welton, *Chem. Rev.*, 1999, **99**, 2071.
4. J. S. Wilkes and M. J. Zaworotko, *J. Chem. Soc., Chem. Commun.*, 1992, 965.
5. J. Fuller, R. T. Carlin, H. C. De Long and D. Haworth, *J. Chem. Soc., Chem. Commun.*, 1994, 299.
6. G. Drake, T. Hawkins, K. Tollison, L. Hall, A. Vij and S. Sobaski, in *Ionic Liquids IIIB: Fundamentals, Progress, Challenges, and Opportunities, Transformations and Processes*, ed. R. D. Rogers and K. R. Seddon, ACS Symposium Series 902, American Chemical Society, Washington DC, 2005, p. 259.
7. A. S. Larsen, J. D. Holbrey, F. S. Tham and C. A. Reed, *J. Am. Chem. Soc.*, 2000, **122**, 7264.
8. P. Wasserscheid, D. Gerhard, S. Himmler, S. Hörmann and P. S. Schulz, in *Ionic Liquids IV: Not Just Solvents Anymore*, ed. J. F. Brennecke, R. D. Rogers and K. R. Seddon, ACS Symposium Series 975, American Chemical Society, Washington DC, 2005, p. 258.
9. R. S. Varma and V. V. Namboordiri, *Chem. Commun.*, 2001, 643.
10. R. S. Varma, in *Ionic Liquids as Green Solvents: Progress and Prospects*, ed. R. D. Rogers and K. R. Seddon, ACS Symposium Series 856, American Chemical Society, Washington DC, 2003, p. 82.
11. J.-M. Lévêque, J.-L. Luche, C. Pétrier, R. Roux and W. Bonrath, *Green Chem.*, 2002, **4**, 357.
12. J. D. Holbrey, W. M. Reichert, R. P. Swatloski, G. A. Broker, W. R. Pitner, K. R. Seddon and R. D. Rogers, *Green Chem.*, 2002, **4**, 407.
13. M. Aresta, I. Tkatchenko and I. Tommasi in *Ionic Liquids as Green Solvents: Progress and Prospects*, ed. R. D. Rogers and K. R. Seddon, ACS Symposium Series 856, American Chemical Society, Washington DC, 2003, p. 93.
14. Z. Fei, T. J. Geldbach, D. Zhao and P. J. Dyson, *Chem. Eur. J.*, 2006, **12**, 2122.
15. A. E. Visser, R. P. Swatloski, W. M. Reichert, R. Mayton, S. Sheff, A. Wierzbicki, J. H. Davis Jr. and R. D. Rogers, *Chem. Commun.*, 2001, 135.

16. K. Fukumoto, M. Yoshizawa and H. Ohno, *J. Am. Chem. Soc.*, 2005, **127**, 2398.
17. S. Himmler, S. Hörmann, R. van Hal, P. S. Schulz and P. Wasserscheid, *Green Chem.*, 2006, **8**, 887.
18. H. Clavier, L. Boulanger, N. Audic, L. Toupet, M. Mauduit and J.-C. Guillemin, *Chem. Commun.*, 2004, 1224.
19. M. Y. Machado and R. Dorta, *Synthesis*, 2005, 2473.

CHAPTER 4

Properties of Ionic Liquids

4.1 INTRODUCTION

The science of room-temperature ionic liquids is a fledgling science. Every month, science journals report the synthesis of novel ionic liquids. The list grows endlessly. Knowledge about the physical, chemical and biological properties of many ionic liquids is limited compared with conventional organic solvents although it is growing at a phenomenal rate.

The determination of the properties of ionic liquids and the trends in these properties is critical in the design of ionic liquids for specific applications. For some properties, such as vapour pressure, it possible to make generalizations that apply to all classes of ionic liquids. For other properties, generalizations that apply to all ionic liquids or specific classes of ionic liquids are not possible at present.

One of the impressive features of ionic liquids is the wide variation in their properties. For example, some are acidic, some basic and others neutral. Some are miscible with water and others immiscible. Some are toxic while others are non-toxic. This variation in properties opens up numerous opportunities for modifying the properties of cations and anions independently. Ionic liquids therefore have a dual nature that can be exploited for various applications. Examples of this wide variation of properties are presented in later chapters.

The sheer number of reported ionic liquids also makes the acquisition and collation of their physical, chemical and biological data a monumental task. Much of the research on ionic liquids has focused on those

An Introduction to Ionic Liquids
By Michael Freemantle
© Michael Freemantle 2010
Published by the Royal Society of Chemistry, www.rsc.org

with imazolium cations and so, inevitably, much of the data that has been acquired so far focuses on these types of ionic liquids.

4.2 LIQUID RANGE AND THERMAL STABILITY

Ionic liquids have far higher liquid ranges than molecular solvents. Liquid range, also known as liquidus range, is the temperature range between melting point or glass transition temperature and boiling point or thermal decomposition temperature.

Many ionic liquids slowly form glasses at low temperatures. Because they have negligible vapour pressures they generally do not evaporate or boil at high temperatures. The upper limit of their liquidus range is therefore determined by the thermal decomposition temperature.

For example, 1-alkyl-3-methylimidazolium salts typically have glass transition temperatures in the range -70 to $-90\,°C$ and thermal decomposition temperatures ranging from 250 to over $450\,°C$. They therefore have liquid ranges of over $300\,°C$. For comparison, water is liquid from 0 to $100\,°C$ at atmospheric pressure and so has a liquid range of $100\,°C$. Ethanol has a melting point of $-114.1\,°C$ and a boiling point of $78.5\,°C$ and, therefore, a liquid range of $192.6\,°C$.

The high thermal decomposition temperatures, *i.e.* the high thermal stabilities, of ionic liquids means that experiments can be carried out in these solvents at high temperatures without any solvent degradation.

4.3 MELTING POINTS

The melting points of room-temperature ionic liquids tend to decrease as the size of the anion or cation increases. For example, the melting point of $[C_2mim]Cl$ is $87\,°C$ whereas that of $[C_2mim][AlCl_4]$, which has a much larger anion, is $7\,°C$. Small variations in the length of the alkyl chain in a cation can also lead to huge differences in the melting points. The melting point of $[C_4mim]Cl$ is $65\,°C$, that is over $20\,°C$ lower than $[C_2mim]Cl$, which has a smaller cation.

The symmetry of the cation also significantly influences melting point. As symmetry increases, the ions pack more efficiently and the melting point of the ionic liquid increases. For example, the tetraalkyl-ammonium bromide $[N_{5\ 5\ 5\ 5}]Br$, which has four straight-chain pentyl groups, has a melting point of $101.3\,°C$. In contrast, $[N_{1\ 5\ 6\ 8}]Br$, which has four different alkyl groups (a methyl group and straight-chain pentyl, hexyl and octyl groups) is liquid at room temperature.

The melting point of an ionic liquid can be lowered by adding another salt to form a eutectic mixture. For example, a eutectic mixture of

[C$_2$mim]Cl and AlCl$_3$ in the mole ratio 1 : 2 has a melting point of $-96\,°C$ compared with $87\,°C$ for pure [C$_2$mim]Cl.

An investigation by Scurto and Leitner revealed that pressurizing simple organic salts with compressed CO$_2$ can lead to remarkably high melting point depressions.[1] The two scientists showed that the melting points of a range of imidazolium, ammonium and phosphonium salts that are normally solid at room temperature are up to $120\,°C$ lower under CO$_2$ pressure than their normal melting points. The findings open up the possibility of bringing solid organic salts that are not normally considered for ionic liquids applications into the realm of room-temperature ionic liquids. For example, the salts might be used as liquid phases in catalytic processes at temperatures far below their regular melting points.

4.4 VAPOUR PRESSURE

In 2005, Rebelo and colleagues noted that no reliable experimental data on the vapour pressures and their dependence on temperature existed for ionic liquids.[2]

The lack of measurable vapour pressure at temperatures up to their thermal decomposition temperatures arises from the strong coulombic interactions between the ions in the liquids. Negligible vapour pressure is one of the most celebrated properties of ionic liquids. Ionic liquids therefore generally do not evaporate in reaction vessels and cannot contribute to air pollution or cause health concerns in this context.

Even so, it is possible to distil certain ionic liquids at high temperature and low pressure. Using experimental surface tension and density data, Rebelo's group predicted that it should be possible to distil ionic liquids with [NTf$_2$]$^-$ anions and imidazolium cations containing long alkyl chain lengths at temperatures between their estimated boiling and decomposition temperatures. They subsequently carried out distillations of [C$_{10}$mim][NTf$_2$] and [C$_{12}$mim][NTf$_2$] at reduced pressure and $70\,°C$.

In 2007, MacFarlane and co-workers described a study of ionic liquids with tetraalkylphosphonium or tetraalkylammonium cations and various anions.[3] They reported that ionic liquids with [P$_{6\ 6\ 6\ 14}$]$^+$ or [N$_{4\ 4\ 4\ 4}$]$^+$ cations and chloride or [NTf$_2$]$^-$ anions exhibit properties that reflect strong ion association. These include relatively low viscosity and "a degree of volatility". They concluded that the liquids are intermediate between true ionic liquids and true molecular solvents.

In general, however, the vapour pressures of ionic liquids, notably the widely-used imidazolium ionic liquids with short cationic alkyl chains, are negligible at ambient temperatures and pressures. Consequently, many ionic liquids show little or no evidence of distillation below their thermal decomposition temperatures.

4.5   HEAT CAPACITY AND HEAT TRANSFER

The design and operation of chemical reactors, pumps, heat-transfer equipment such as refrigeration systems, and other industrial equipment requires knowledge of a range of thermal properties of the fluids used in the processes. These properties include melting points, thermal stabilities, boiling points and heat capacities.

The specific heat capacity of a substance is the energy required to raise the temperature of a unit mass of the substance by one kelvin. Liquids generally have specific heat capacities between 1.6 and 2.1 $J\,g^{-1}\,K^{-1}$. Hydrogen-bonding liquids have higher values. For example, the specific heat capacity of water is $4.186\,J\,g^{-1}\,K^{-1}$.

In 2003, Holbrey and co-workers reported the specific heat capacities of five widely used ionic liquids with imidazolium cations: [C$_4$mim]Cl, [C$_4$mim][PF$_6$], [C$_2$mim][PF$_6$], [C$_6$mim][PF$_6$] and [C$_4$mim][NTf$_2$].[4] They found that the specific heat capacities ranged from 1.17 to $1.80\,J\,g^{-1}\,K^{-1}$ at 100 °C and increased linearly with temperature. These values are comparable to those of fluids used for heat transfer applications.

Wilkes and co-workers have suggested that ionic liquids might be suitable as heat-transfer fluids for large-scale solar energy collectors to be used for electric power generation.[5] The group determined the heat capacities and other thermal properties of three ionic liquids: [C$_2$mim][BF$_4$], [C$_4$mim][BF$_4$] and 1,2-dimethyl-3-propylimidazolium bistriflamide. They concluded that the ionic liquids are superior in many ways to existing commercial heat-transfer fluids. For example, they are stable over a wide range of temperature, have low vapour pressures and can store large amounts of heat.

4.6   VISCOSITY

Viscosity is a measure of a liquid's resistance to flow. Liquids with lower viscosity flow more readily. The centipoise (cP) is commonly used as the physical unit of viscosity.

In general, ionic liquids are more viscous than molecular solvents. The viscosities of ionic liquids at room temperature typically lie in the range of 10 to over 500 cP.[6] For example, the viscosities of [C$_2$mim][BF$_4$] and [C$_4$mim][PF$_6$] at 25 °C are 34 and 270 cP, respectively. [C$_2$mim]Cl-AlCl$_3$, with a 1 : 1 mole ratio of the two components, has a viscosity of 18 cP. In comparison, acetone, water and ethanol have viscosities of 0.31, 0.89 and 1.07 cP, respectively, at 25 °C. The room temperature viscosity of ethylene glycol is 16.1 cP and that for glycerol is 934 cP.

The viscosities of ionic liquids generally increase with increasing size of the cation, and particularly with increasing alkyl chain lengths. For example, the viscosities of $[N_{6\,2\,2\,2}][NTf_2]$ and $[N_{8\,2\,2\,2}][NTf_2]$ at 25 °C are 167 and 202 cP, respectively.

Ionic liquids with weakly coordinating anions, such as $[BF_4]^-$, $[PF_6]^-$ and $[NTf_2]^-$, have lower viscosities than those with strongly coordinating anions. For example, the room temperature viscosities of $[C_6mim][BF_4]$ and $[C_6mim][NO_3]$ are 314 and 804 cP, respectively.

As expected, viscosities of ionic liquids decrease with increasing temperature. The changes can be dramatic. For example, the viscosity of $[C_4mim][PF_6]$ decreases by almost 30% when the temperature is increased from 20 to 25 °C.

Impurities can also have a dramatic impact on ionic liquid viscosities (Section 4.15).

4.7 DENSITY

Ionic liquids are generally denser than water. Consequently, if an ionic liquid does not mix with water it forms the lower phase when the two liquids are mixed.

Ionic liquids with shorter alkyl chains or less bulky cations have higher densities than ionic liquids with longer chains or more bulky cations. For haloaluminate ionic liquids, densities increase with increasing proportions of the aluminium halide in the binary mixture.

In contrast to viscosity, changes of temperature have minimal impact on the density of an ionic liquid.

4.8 SOLUBILITY AND MISCIBILITY

The solubility, miscibility and immiscibility properties of ionic liquids vary widely, depending on the nature of the cations and anions. Chapter 5 describes these characteristics in detail.

4.9 WATER STABILITY

The stability of ionic liquids in water also varies widely. Many are stable in the presence of water. But there are some notable exceptions.

Haloaluminate-based ionic liquids, such as [C$_4$mim]Cl-AlCl$_3$, react vigorously with water, or moisture. They decompose to form hydrogen halide gas and corrosive hydrohalic acid. They should therefore be handled in a dry-box.

Ionic liquids with [PF$_6$]$^-$ and [BF$_4$]$^-$ cations are also unstable in the presence of water. The anions hydrolyse and liberate hydrogen fluoride (Section 6.2)

4.10   CONDUCTIVITY

Ionic liquids, by definition, exhibit ionic conductivity, *i.e.* electrical conductivity (also known as specific conductivity). Conductivity has units of siemens per metre (S m^{-1}) or, for low conductivities, millisiemens per centimetre (mS cm^{-1}). The conductivity of an ionic liquid is a measure of the liquid's ability to conduct an electric current. The ions act as charge carriers.

The conductivity of ionic liquids is discussed in Section 7.1.

4.11   ELECTROCHEMICAL POTENTIAL WINDOW

The electrochemical potential window is the voltage range over which a material is neither reduced or oxidized at an electrode and is therefore electrochemically inert. For an ionic liquid, the window depends principally on the cation's resistance to reduction and the anion's resistance to oxidation.

In general, non-haloaluminate room temperature ionic liquids have wide electrochemical windows, typically in the range 2.0–6.0 V (Section 7.1). Impurities, such as halide ions, can have a significant impact on the size of these windows.

4.12   SURFACE TENSION

There have been relatively few reports of measurements of the surface tensions of ionic liquids. Most have focused on imidazolium ionic liquids.

One of the first studies was published by Law and Watson in 2001.[7] They measured the surface tensions of eight ionic liquids, consisting of [C$_4$mim]$^+$, [C$_8$mim]$^+$ or [C$_{12}$mim]$^+$ cations paired with [PF$_6$]$^-$, [BF$_4$]$^-$, Cl$^-$ or Br$^-$ anions over a range of temperatures.

The two chemists showed that the surface tension of all these ionic liquids decrease linearly with temperature. For ionic liquids containing the same anion, the surface tension at a specific temperature decreases with increasing alkyl chain length in the imidazolium cation. For ionic liquids with the same cation, the surface tension generally increases with increasing size of the anion.

In general, the surface tensions of ionic liquids are modest and comparable to those of organic liquids.[8] Ionic liquid surface tensions are also lower than those of water and molten inorganic salts.

4.13 FLUORESCENCE

As with other physicochemical properties, studies of the optical properties of ionic liquids have focused principally on those compounds with imidazolium cations.

Imidazolium ionic liquids are generally considered to be completely optically transparent in the visible region and in much of the UV region. Several studies suggest that any electronic absorption and fluorescence emission detected in these regions is caused by impurities in the ionic liquids.[9,10]

In 2005, Samanta and co-workers described an investigation into the optical properties of neat [C_2mim][BF_4] and [C_4mim][BF_4].[11] The study showed that the two ionic liquids are characterized by non-negligible absorption in the entire UV region with a long absorption tail extending into the visible region. They attributed this absorption to the imidazolium cations. Furthermore, when the ionic liquids were excited in these regions, they exhibited fluorescence over a significant portion of the visible region. Fluorescence spectra revealed that the peak fluorescence emission wavelengths depended on the excitation wavelength.

The authors concluded from their results that the two ionic liquids "may have serious drawbacks" in some optical studies, particular those involving weakly fluorescent samples. They suggested that ionic liquids with pyrrolidinium cations might be more suited for optical studies because they have saturated rings.

4.14 REFRACTIVE INDICES

The refractive index of a substance is a measure of its ability to refract light when it travels from the substance into another medium. Substances with a refractive greater than 1.6 are generally regarded to have a high refractive index.

Various studies have shown that the refractive index of an ionic liquid depends on the nature of both the cation and anion. For example, refractive index increases as the number, branching and length of alkyl chains in the cation increases.

In 2005, Deetlefs and co-workers reported the synthesis of a range of ionic liquids with 1-alkyl-3-methylimidazolium cations and polyhalide anions such as [C_2mim][I_5] and [C_4mim]BrI_2] and measurements of their refractive indices.[12] The compounds exhibited refractive indices between 1.68 and 2.23. Such compounds could be potentially useful as immersion fluids for examining gems and minerals (Section 14.9).

4.15   IMPACT OF IMPURITIES

Impurities in ionic liquids – such as traces of water, acids, halide ions, residual solvents and unreacted volatile organic compounds arising from the preparation of the liquids – can have a pronounced impact on the physical, spectroscopic and chemical characteristics of the liquids.

In 2000, Seddon and co-workers described a systematic study of the effects of impurities and additives, such as water, chloride and co-solvents, on the physical properties of ionic liquids.[13] They noted, for example, that the melting points reported in various earlier studies for the ionic liquid [C_2mim][BF_4] ranged from 5.8 to 15 °C. The values clearly varied depending on impurities in the ionic liquid. An even more striking example is 1,2-dimethyl-3-propylimidazolium chloride, the reported melting point of which varies from 58–66 to 138 °C, depending on the report.[14]

The viscosities of [C_2mim][BF_4] and other 1-methylimidazolium salts have also been shown in increase dramatically when small amounts of chloride impurities are present. Halide impurities also deactivate transition metal-based catalysts immobilized in ionic liquids.

Halide impurities typically occur in ionic liquids prepared in aqueous solutions by anion metathesis of an organic halide with an inorganic salt. Unlike conventional molecular organic solvents, most ionic liquids cannot be purified by distillation as they have negligible vapour pressure. Various other methods are therefore used to lower the halide content of ionic liquids. They include the use of silver nitrate to precipitate the silver halide. Ion chromatography is another commonly used purification method.

The problem can be also be minimized by extracting an ionic liquid into an organic solvent and washing the solution with small amounts of deionized water. The solvent is then removed in a rotary evaporator. The ionic liquid is finally dried by heating under vacuum. This purification method, however, is not suitable for water miscible ionic liquids, such as [C_2mim][BF_4]. In such cases, halide-free ionic liquids are prepared using an ion-exchange resin or by alkylation with a halide-free alkylating agent.

As many ionic liquids do not crystallize when cooled, they cannot be purified by crystallization.

Ionic liquids often have a yellow to brown discoloration caused by trace amounts of impurities. The colour of an ionic liquid is typically removed by stirring the liquid with activated charcoal and then passing it through a column of alumina.

Even so, the use of sorbents such as activated charcoal, alumina and silica to remove organic contaminants can itself result in contamination. In 2008, Clare and co-workers showed that ionic liquids exposed to alumina or silica sorbents are contaminated with parts-per-million levels of sorbent particles even after nanofiltration.[15] Trace sorbents, the authors noted, are likely to have a non-negligible impact on the electrochemical, spectroscopic and catalytic properties of ionic liquids.

REFERENCES

1. A. M. Scurto and W. Leitner, *Chem. Commun.*, 2006, 3681.
2. L. P. N. Rebelo, J. N. C. Lopes, J. M. S. S. Esperança and E. Filipe, *J. Phys. Chem. B*, 2005, **109**, 6040.
3. K. J. Fraser, E. I. Izgorodina, M. Forsyth, J. L. Scott and D. R. MacFarlane, *Chem. Commun.*, 2007, 3817.
4. J. D. Holbrey, W. M. Reichert, R. G. Reddy and R. D. Rogers, in *Ionic Liquids as Green Solvents: Progress and Prospects*, ed. R. D. Rogers and K. R. Seddon, ACS Symposium Series 856, American Chemical Society, Washington DC, 2003, p. 121.
5. Van Valkenburg, R. L. Vaughn, M. Williams and J. S. Wilkes, *Thermochim. Acta*, 2005, **425**, 181.
6. J. D. Holbrey and R. D. Rogers, in *Ionic Liquids in Synthesis*, ed. P. Wasserscheid and T. Welton, Wiley-VCH, Weinheim, 2008, **vol. 1**, p. 57.
7. G. Law and P. R. Watson, *Langmuir*, 2001, **7**, 6138.
8. W. Martino, J. F. de la Mora, Y. Yoshida, G. Saito and J. Wilkes, *Green Chem.*, 2006, **8**, 390.
9. A. Pau, P. K. Mandal and A. Samanta, *Chem. Phys. Lett.*, 2005, **402**, 375.
10. M. Koel, in *Ionic Liquids in Chemical Analysis*, ed. M. Koel, CRC Press, Boca Raton, FL, 2009, p. 295.
11. A. Paul, P. K. Mandal and A. Samanta, *J. Phys. Chem. B*, 2005, **109**, 9148.

12. M. Deetlefs, M. Shara and K. R. Seddon, in *Ionic Liquids IIIA: Fundamentals, Progress, Challenges, and Opportunities*, ed. R. D. Rogers and K. R. Seddon, ACS Symposium Series 901, American Chemical Society, Washington DC, 2005, p. 219.
13. K. R. Seddon, A. Stark and M.-J. Torres, *Pure Appl. Chem.*, 2000, **72**, 2275.
14. M. Klingele, *Ionic Liquids Today*, 2008, **1**, 3.
15. B. R. Clare, P. M. Bayley, A. S. Best, M. Forsyth and D. R. MacFarlane, *Chem. Commun.*, 2008, 2689.

CHAPTER 5

Ionic Liquids as Designer Solvents

5.1 INTRODUCTION

During the 1980s, much of the research on the use of ionic liquids as solvents for chemical reactions focused on chloroaluminate salts with pyridinium and imidazolium cations. However, the anions of these liquids hydrolyse in the presence of moisture, forming strongly acidic and corrosive hydrogen chloride gas. The reactivity of these ionic liquids with water therefore severely limits their usefulness as solvents for organic reactions.

In 1992, the discovery of room-temperature ionic liquids with per-fluorinated anions, such as $[PF_6]^-$ and $[BF_4]^-$, transformed research on ionic liquids. At the time, the liquids were thought to be both air and water stable and therefore ideal replacements for volatile organic solvents. As a result, much of the subsequent research on chemistry in ionic liquids employed these salts as solvents. However, it is now known that the anions hydrolyse under acidic conditions to form hydrogen fluoride. The ionic liquids are therefore not environmentally benign and not ideal replacements for organic molecular solvents.

The concept of ionic liquids as "designer solvents" attracted a lot of attention when the term was first introduced in the late 1990s and has driven much of the research since then. Such designer solvents were seen as potential "drop-in" replacements for the environmentally-hazardous volatile organic solvents that are widely used in the chemical industry. In practice, however, ionic liquids have usually been selected rather than designed for specific applications. The selection has often been based on

An Introduction to Ionic Liquids
By Michael Freemantle
© Michael Freemantle 2010
Published by the Royal Society of Chemistry, www.rsc.org

trial and error, and almost certainly dominated by expedience. Much of the recent research literature on ionic liquids has focused on ionic liquids that are readily synthesized by well-established methods and, in an increasing number of cases, are commercially available.

In theory at least, ionic liquids can be designed to deliver almost any set of physical and chemical properties for almost any application in the chemical sciences. The solubility and miscibility characteristics of ionic liquids can therefore, in principle, be tailored for specific applications by changing the structure and nature of the cations and/or anions. The tailoring, however, requires knowledge of patterns and trends in the solubility and miscibility characteristics of ionic liquids. A wealth of data has accumulated on some families of ionic liquids, particularly the imidazolium family. For such families, certain trends are apparent. But solubility data on other families of ionic liquids are limited. It is therefore difficult or risky to draw generalizations for ionic liquids as a whole.

5.2 SOLUBILITY AND MISCIBILITY

One of the exciting features of room-temperature ionic liquids is the broad variation of their solubility and miscibility properties. They can, for example, dissolve both ionic and covalent compounds. This feature makes ionic liquids attractive not only as solvents for chemical processes but also for the separation and extraction of materials from solutions and mixtures. In addition, it is well established that solvents can have a pronounced impact on the course of a chemical reaction. The development of ionic liquids as solvents for various materials and chemical processes has therefore opened up a new face of chemistry: the study of chemistry in an ionic environment rather than in a molecular environment.

Seddon's discovery in the early 1980s that ionic liquids could dissolve kerogen is a classic example of the power of ionic liquids to dissolve materials.[1] Kerogen, a fossilized organic material present in sedimentary rocks, was previously found to be insoluble in all known solvents except hydrofluoric acid. The use of this hazardous acid made analysis of the material difficult. Seddon's group showed that kerogen dissolves in chloroaluminate ionic liquids under microwave irradiation. The discovery enabled the kerogen fractions to be separated and analysed, and the rocks to be dated.

Ionic liquids can be selected or designed to dissolve a wide range of organic and inorganic gases, liquids and solids. For example, carbon dioxide, benzene, carbohydrates, antibiotics, coal and rocks are all soluble in certain types of ionic liquid.

The ability of an ionic liquid to dissolve a substance depends on several factors, most notably its polarity and the coordination ability of its ions. The coordination ability of ionic liquid anions, for example, influences the solubility of metal salts in ionic liquids.

5.3 POLARITY AND SOLVATION

Ionic liquids are generally viewed as polar substances because they consist of ions and are therefore electrically charged. As a rule, polar solvents, such as water, are better than organic solvents at dissolving charged solutes, like metal salts. However, ionic liquids can dissolve both polar and non-polar solutes. In general, ionic liquids tend to be more strongly solvating than conventional organic molecular solvents.

The key factor that determines the overall ability of any solvent, whether molecular or ionic, to dissolve solutes is the polarity of the solvent. Solvent polarity, however, is a complex and somewhat nebulous concept. It has been defined as the "overall solvation capability (or solvation power)" of the solvent. Solvation is the association of solvent molecules, or ions in the case of ionic liquids, with solute ions or molecules. Solvation in ionic liquids can occur through various solute–solvent interactions, including ionic interactions, dipole interactions, hydrogen bonding, van der Waals forces and aromatic interactions. The interactions that come into play on dissolution largely depend on the nature of the anions and cations in the ionic liquid and also on the nature of the solute.

Several methods have been used to determine solvent polarity. One relative measure of the polarity of a liquid is its dielectric constant, also known as relative permittivity. It is a macroscopic physical parameter that depends, in the case of molecular solvents, on the properties of the molecules that make up the liquid. Liquids consisting of non-polar molecules, such as benzene, have low dielectric constants, whereas those with polar molecules, such as nitrobenzene, have relatively high dielectric constants. Hydrogen-bonded liquids, such as water, have exceptionally high dielectric constants.

Conventional methods of determining dielectric constants require a non-conducting medium. Since ionic liquids consist of ions rather than molecules, they conduct electricity. Conventional methods therefore cannot be used to determine the dielectric constants of ionic liquids. However, the constants can be determined indirectly using a method known as microwave dielectric spectroscopy.

Other methods have also been employed to characterize the solvent polarity of ionic liquids. They include several that rely on spectroscopic techniques and polarity-sensitive dyes. One empirical polarity scale, based on such a technique, puts the polarity of the non-polar tetra-methylsilane at 0.00 and that for water at 1.00.[2] Ionic liquid polarities determined by this method lie in the range 0.35 to 1.10. For example, imidazolium ionic liquids have polarities between 0.50 and 0.75, which is similar to the range for alcohols such as methanol and ethanol. For comparison, the polarities of benzene, dimethylformamide and dimethyl sulfoxide are 0.10, 0.40 and 0.45, respectively.

5.4 MISCIBILITY WITH WATER

The miscibility of ionic liquids with water is of particular interest since hydrophobic ionic liquids can be used as water-immiscible polar phases in biphasic processes.

The coordinating ability of the ions in an ionic liquid is one of the key factors in determining an ionic liquid's miscibility with water. Ionic liquids with basic anions, such as Cl^- and $[NO_3]^-$, are strongly coordinating, whereas those with acidic anions, such as $[Al_2Cl_7]^-$, are non-coordinating. $[BF_4]^-$, $[PF_6]^-$ and $[NTf_2]^-$ are weakly coordinating anions. In general, as the coordinating ability of the anion decreases the hydrophobicity of the ionic liquid increases. $[C_4mim][PF_6]$, for example, is hydrophobic, whereas $[C_4mim]Cl$ is hydrophilic.

Cations can also be used to control the hydrophobicity of ionic liquids. As the length of the alkyl chain on cations such as pyridinium and imidazolium increases, the ionic liquids tend to become less miscible with water. For example, the hydrophobicity of the $[C_nmim][BF_4]$ series of ionic liquids increases as the alkyl chain length increases. In this series, ionic liquids with $n = 2–5$ are miscible with water at room temperature whereas those with $n > 5$ are not.

These differences illustrate that even a minor change in the structure of a cation or anion can have a significant impact on the solubility characteristics of a family of ionic liquids.

As expected, the miscibility of ionic liquids with water increases with temperature. For example, $[C_4mim][BF_4]$ is miscible with water at room temperature but forms a separate phase at temperatures lower than $5\,°C$.

5.5 MISCIBILITY WITH ORGANIC COMPOUNDS

Ionic liquids can, in principle, be selected or designed to be miscible or immiscible not only with water but also with all organic solvents. Much depends on the polarities of the ionic liquids. These tend to be higher than those of chlorinated organic solvents, such as dichloromethane, but lower than that of water. Ionic liquids are generally miscible with liquids with medium to high dielectric constants, *i.e.* liquids consisting of polar molecules. On the other hand, they tend to be immiscible with liquids with low dielectric constants. For example, [C$_4$mim][BF$_4$] is miscible with methanol and dichloromethane, both of which have relatively high dielectric constants, but immiscible with toluene, which has a much lower dielectric constant. Table 5.1 illustrates the variation in miscibilities and solubilities of various liquid and solid organic solutes in [C$_4$mim][BF$_4$].[3]

In general, polar compounds tend to be more soluble in ionic liquids than apolar compounds. For example, unsaturated organic compounds are more polar and more soluble in ionic liquids than saturated organic compounds, notably the alkanes which are generally immiscible with ionic liquids. As a result, alkanes can be used in biphasic systems consisting of an organic phase and a polar ionic liquid phase. In ionic liquid/aqueous biphasic systems, on the other hand, the ionic liquid is the less polar phase.

For a given ionic liquid, the solubility of alkenes decreases as the length of their alkyl chains, and therefore their non-polar character, increases. In addition, for a given alkene, the solubility in a family of ionic liquids increases as the length of alkyl chain on the ionic liquid cation increases. For example, the solubility of oct-1-ene in methyltrialkylammonium tosylate ionic liquids increases as the length of the alkyl chains increase.[4]

Intriguingly, benzene, which consists of non-polar molecules, is soluble up to 50 vol.% in chloroaluminate ionic liquids. Several studies

Table 5.1 Solubilities of organic solutes in [C$_4$mim][BF$_4$] at 22 °C. (Data from Blanchard *et al.*[3]).

Solute	Solubility (mole fraction of solute)
Hexane	Immiscible
Benzamide (solid)	0.04
Butyl ethyl ether	0.06
Cyclohexane	0.21
Hexyl alcohol	0.26
Benzene	0.66
Phenol (solid)	0.69
Aniline	Miscible
Benzaldehyde	Miscible
Acetophenone	Miscible

have shown that, in general, benzene and other aromatic hydrocarbon liquids, such as toluene and the xylenes, are soluble in, but rarely completely miscible with, ionic liquids. This phenomenon has been attributed to the formation of liquid clathrates, *i.e.* cage-like structures in the liquid.[5] The clathrates typically consist of ionic liquid anion–cation cages that surround the aromatic molecules.

Ionic liquids can also dissolve natural and synthetic polymers. An example is cellulose, which has been shown in various studies to dissolve in ionic liquids such as [C$_2$mim]Cl and the corresponding acetate [C$_2$mim][OAc] (Section 12.2).

5.6 SOLUBILITY OF METAL SALTS

Metal salts tend to be poorly soluble in ionic liquids with non-coordinating anions, such as fluorinated anions. They are more soluble in ionic liquids with coordinating anions such as chlorides and nitrates, although the solubility tends to be highly variable. LiCl, for example, is very soluble in [C$_2$mim]Cl whereas KCl and NaCl are virtually insoluble.

Metal salts that are soluble typically form binary ionic liquids containing anionic complexes of the metal and the coordinating anions of the original organic salts. For example, AlCl$_3$ dissolves in excess [C$_2$mim]Cl to form the binary ionic liquid [C$_2$mim]Cl-AlCl$_3$, which contains the chloride and chloroaluminate, [AlCl$_4$]$^-$, anions.

One of the advantages of the binary nature of haloaluminate ionic liquids is the opportunity to adjust their Lewis acidity by varying the molar ratio of the organic salt and the metal halide. In 1988, Hussey noted that metal halide compounds generally dissolve in basic ionic liquids to form halometallate ionic liquids with well-defined anionic complexes.[6] He observed that binary room-temperature haloaluminate ionic liquids, such as [C$_4$py]Cl-AlCl$_3$, [C$_2$mim]Cl-AlCl$_3$ and [C$_2$mim]Br-AlBr$_3$, are "excellent solvents in which to study the solution chemistry of transition metal halide complexes and related species." This property has been exploited in transition metal catalysis since it allows ionic liquids containing transition metal complex anions to be used as catalysts and solvents simultaneously.

5.7 GAS SOLUBILITY

Knowledge of the solubility of gases in ionic liquids is important for several reasons. For example, the use of an ionic liquid as a solvent for a reaction such as an oxygenation or hydrogenation with oxygen or hydrogen gas requires that the gas dissolves in the ionic liquid. Ionic liquids have also been shown to be useful as storage media for gases and for separating a gas from a mixture of gases.

In 2001, Brennecke and co-workers established that water vapour dissolves in three water-stable imidazolium ionic liquids: $[C_4mim][PF_6]$, $[C_8mim][PF_6]$ and $[C_8mim][PF_6]$.[7] The following year, the same team reported the solubilities of nine different gases in $[C_4mim][PF_6]$: carbon dioxide, ethylene, ethane, methane, argon, oxygen, carbon monoxide, hydrogen and nitrogen.[8] They found CO_2 to have the highest solubility followed by $CH_2=CH_2$ and C_2H_6. The solubilities of Ar and O_2 were low and those of CO, H_2, and N_2 were negligible.

In 2006, researchers in France published experimental values for the solubility of CO_2, C_2H_6, CH_4, O_2, N_2, H_2, Ar and CO in $[C_4mim][BF_4]$ at atmospheric pressure.[9] The work revealed that CO_2 is the most soluble gas in this ionic liquid. C_2H_6 and CH_4 are less soluble than CO_2 but more soluble than the other five gases. H_2 is the least soluble of the eight gases.

As might be expected, the solubility of gases in ionic liquids increases with increasing pressure. The solubility of gases, such as CO_2, that are highly soluble in ionic liquids decreases with increasing temperature. In contrast, H_2, which is one of the least soluble gases in ionic liquids, is more soluble at higher temperatures.[10] Finally, changes in temperature have little influence on the solubility of sparsely soluble gases like O_2 and CO.

The nature of the cations and anions in the ionic liquid plays a key role in determining the solubility of a gas. A study of the low-pressure solubility of CO_2 in a series of imidazolium-based room-temperature ionic liquids illustrates the point.[11] The study showed that CO_2 is more soluble in an ionic liquid with the fluorinated cation, $[C_8F_{13}mim]^+$, compared with its non-fluorinated counterpart, $[C_8mim]^+$. CO_2 solubility also increases with the length of the alkyl side chain on the imidazolium ring. For example, CO_2 is more soluble in $[C_8mim][Tf_2N]$ than in $[C_3mim][Tf_2N]$. The study also revealed that anions can influence CO_2 solubility: the gas is more soluble in $[C_3mim][NTf_2]$ than in $[C_3mim][PF_6]$.

Brennecke and co-workers have also investigated the solubility of CO_2 in a range of imidazolium ionic liquids.[12] Their results suggest that the anion dominates interactions with CO_2, with the cation playing a secondary role. The work showed that, of all the anions studied, $[NTf_2]^-$ has the greatest affinity for CO_2. There is little difference in CO_2 solubility between ionic liquids with the $[BF_4]^-$ or $[PF_6]^-$ anions. The researchers concluded that CO_2 organizes strongly about these ions.

Brennecke concluded, from their studies of gas solubilities in ionic liquids, that the wide variety of ionic liquid cations, anions and substituents make the possibility of choosing or tailoring ionic liquids for specific gas separations or reactions involving gases "an exciting option."[13]

Interestingly, the presence of one gas can influence the solubility of another in an ionic liquid. For example, Aki and others have demonstrated that CO_2 enhances the solubility of O_2 and CH_4 in [C_6mim][NTf_2], even at low pressures.[14]

5.8   BIPHASIC SYSTEMS

If two liquids form two phases, *i.e.* two layers, when mixed, they are said to be immiscible. The liquid layers in a two-layer or biphasic system are not necessarily pure, however. Each liquid may have some of the other liquid dissolved in it. A binary ionic liquid/solvent system is therefore likely to consist of an ionic liquid-rich phase and a solvent-rich phase.

In principle, ionic liquids can be selected or designed to be immiscible with a wide range of organic solvents and can therefore provide a polar alternative to water for biphasic systems consisting of a liquid organic phase and a liquid polar phase. Hydrophobic ionic liquids can also be used as immiscible polar phases with water. However, hydrophobic ionic liquids almost invariably become "wet" when mixed with water. In such a biphasic system, the bottom hydrophobic ionic liquid layer will contain a small amount of water and the top water phase will contain some dissolved ionic liquid.

The development of ionic liquids that are immiscible with reactants and products but dissolve catalysts has been of particular interest over the past decade or so. These biphasic catalytic processes offer the benefits of both homogeneous and heterogeneous catalysis. For example, they generally operate under the mild conditions and exhibit the high efficiency and selectivity characteristic of homogeneous catalysis. They also allow the catalyst to be easily separated from the reaction products and re-used – a characteristic of heterogeneous catalysis.

5.9   PHASE DIAGRAMS

The phase diagrams of various ionic liquid systems, notably two-component systems, have been published over the past few decades. Much of the attention has focused on the phase behaviour of ionic liquids that are binary mixtures or on systems where one component is an ionic liquid and the other is a solvent or a gas.

A classic and widely-cited example of a phase diagram of a binary ionic liquid was published in 1984 by Wilkes and co-workers.[15] The diagram depicts the phase behaviour of a two-component chloro-aluminate system consisting of the organic salt [C$_2$mim]Cl and AlCl$_3$ as the mole fraction of AlCl$_3$ increases (Figure 5.1). The melting and freezing points and glass transition temperatures of the [C$_2$mim]Cl-AlCl$_3$ mixture decrease as the mole fraction of AlCl$_3$ increases. When the molar ratio of the organic salt to AlCl$_3$ is 1:2, the melting and freezing points begin to rise. A maximum occurs when the ratio is 1:1. The melting and freezing points then fall to a minimum corresponding to a ratio of 2:1. When the mole fraction of AlCl$_3$ is less than 0.5, the ionic liquid exhibits basicity because of the excess of chloride ions from the organic salt. When the mole fraction is greater than 0.5, the ionic liquid exhibits Lewis acidity. At 0.5, the ionic liquid is neutral.

Rebelo and co-workers have observed that the type of phase diagram shown in Figure 5.2 is commonly encountered for commonly binary ionic liquid/solvent mixtures where the solvent is water or an alcohol.[16] At low temperatures, the two liquids exhibit partial mutual solubility over a range of compositions. There are two phases in this region. Above a certain temperature, there is total mutual miscibility over the complete range of liquid–liquid compositions. This temperature is known as the liquid–liquid critical temperature or upper critical solution temperature (UCST). For some systems, however, the solvent evaporates or the ionic liquid decomposes below this temperature.

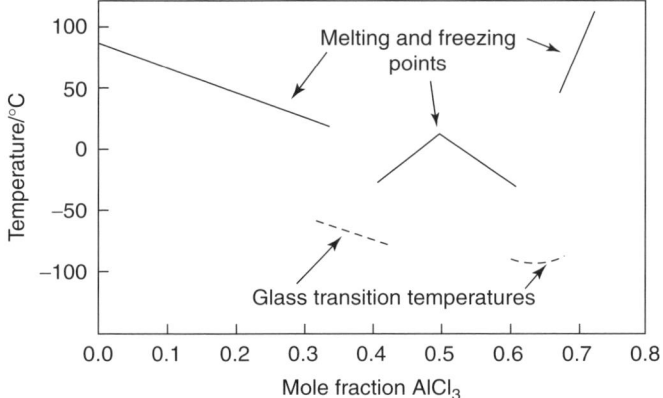

Figure 5.1   Phase diagram of [C$_2$mim]Cl-AlCl$_3$. (Adapted from Wilkes.[15])

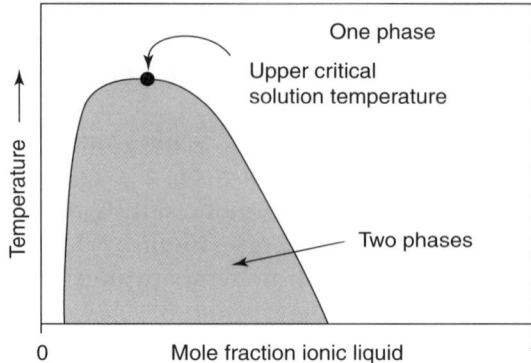

Figure 5.2   Typical phase diagram for a binary ionic liquid/solvent mixture.

An early example of the UCST-type phase diagram was published by Dupont and colleagues in 1998 in a paper that focused on catalysis by palladium compounds immobilized in ionic liquids.[17] The diagram shows that the UCST of a $[C_4mim][BF_4]$/water system is 5 °C. The palladium compounds could therefore be immobilized in the ionic liquid rich phase and the catalytic reactions carried out at temperatures below 5 °C. The mutual solubilities of the two liquids are quite large, however. At 0 °C, the $[C_4mim][BF_4]$-rich phase contains about 27 wt% water and the water phase about 25 wt% $[C_4mim][BF_4]$.

In 2005, Rebelo and colleagues reported the discovery of ionic liquid–organic solvent systems with a rare type of phase diagram.[16] A system consisting of $[C_{4.3}mim][NTf_2]$ and chloroform has two lower critical solution temperatures (LCSTs) as well as an UCST. $[C_{4.3}mim][NTf_2]$ is a mixture of $[C_4mim][NTf_2]$ and $[C_5mim][NTf_2]$. The phase diagram (Figure 5.3) has an upper two-phase region with a LCST and a lower closed-loop two-phase region with an LCST and an UCST.

A binary system consisting of $[C_5mim][NTf_2]$ and a mixture of $CHCl_3$ and carbon tetrachloride (CCl_4) exhibits the phase behaviour shown in Figure 5.4. It also has two so-called "de-mixing" regions with a LCST at a high temperature and a UCST at a lower temperature. By manipulating the alkyl chain length on the imidazolium cation, both systems evolve into an "hour-glass" type of phase diagram. In such a diagram, the UCST merges with the LCST at the higher temperature.

Brennecke and her group have investigated the phase behaviour of ionic liquid/CO_2 systems over a range of pressures.[3] They established that $[C_4mim][PF_6]$/CO_2 is a two-phase system (Figure 5.5). They were unable to detect any $[C_4mim][PF_6]$ in the CO_2-rich phase. This phase is, therefore, essentially pure CO_2. The researchers pointed out that this type of phase behaviour is not found with mixtures of CO_2 and

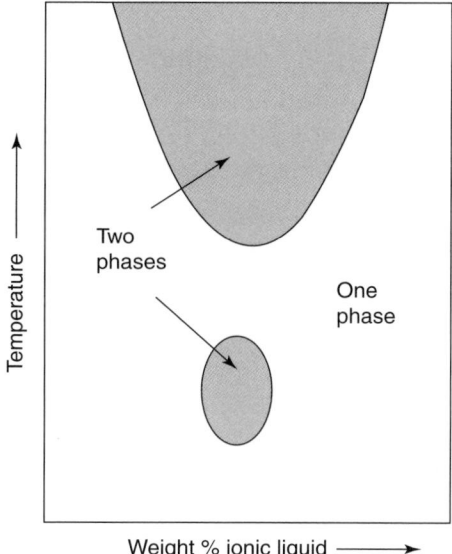

Figure 5.3 The [C$_{4.3}$mim][NTf$_2$]/CHCl$_3$ system exhibits two lower critical solution temperatures and one upper critical solution temperature. (Adapted from Rebelo.[16])

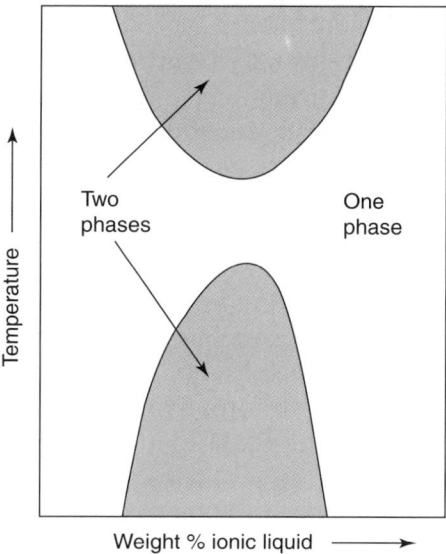

Figure 5.4 The [C$_5$mim][NTf$_2$]/CHCl$_3$ + CCl$_4$ system exhibits one lower critical solution temperature and one upper critical solution temperature. (Adapted from Rebelo.[16])

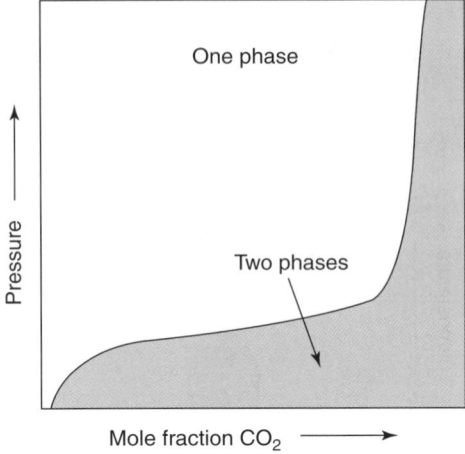

Figure 5.5 Phase diagram of the [C₄mim][PF₆]/CO₂ system. (Adapted from Brennecke.[3])

conventional organic liquids, where significant solubility of the solvent in the CO_2 phase occurs.

5.10 PARTITIONING OF SOLUTES BETWEEN IONIC LIQUIDS AND WATER

Water-immiscible ionic liquids have been considered as potential "drop-in" replacements for the conventional organic molecular solvents that are used to extract solutes from aqueous solutions.

In 1998, Rogers and co-workers described an investigation of the partitioning of benzene derivatives, such as aniline, benzoic acid, chlorobenzene and toluene, between [C₄mim][PF₆] and water.[18] The work showed that the distribution of the solutes between the ionic liquid and water corresponds approximately to the distribution of the same solutes between traditional organic molecular solvents and water. The results indicate that solutes such as aniline or benzoic acid, which have charged groups or the ability to form hydrogen bonds, exhibit a greater preference for the ionic liquid phase than uncharged or non-polar solutes such as benzene.

It is well known that the partitioning of acids and bases, like benzoic acid and aniline respectively, in organic solvent/aqueous systems can be varied by changing the pH of the aqueous solution. Shifting the pH enables the solute to be switched from one phase to the other. In 2000, Rogers' team showed that such switching is also possible for ionic liquid–water systems.[19] The team reversibly switched the indicator dye thymol blue between imidazolium ionic liquids, such as [C$_4$mim][PF$_6$], and water phases by adjusting the pH of the aqueous layer using CO$_2$ gas and NH$_3$ gas. At low pH, thymol blue is red and prefers the ionic liquid phase. At high pH, the dye is blue and prefers the aqueous phase.

5.11 BIPHASIC SYSTEMS OF HYDROPHILIC IONIC LIQUIDS AND WATER

It would seem logical that a hydrophobic, *i.e.* water-hating or water-immiscible, ionic liquid such as [C$_4$mim][PF$_6$] is required to form a biphasic ionic liquid–aqueous system. However, research reported in 2003 revealed that water-miscible ionic liquids such as [C$_4$mim]Cl can be induced to form biphasic systems with aqueous solutions.[20]

The trick is to add a kosmotropic salt such as K$_3$PO$_4$ to an aqueous solution of the ionic liquid (kosmotropic means water-structuring). Water molecules surround the water-structuring phosphate anions, causing the hydrophobic cation [C$_4$mim]$^+$ and the less water-structuring anion Cl$^-$ of the ionic liquid to be salted out. The process affords a biphasic system with a [C$_4$mim]Cl-rich upper phase and a K$_3$PO$_4$-rich lower phase. Other kosmotropic salts, such as K$_2$CO$_3$ and Na$_2$S$_2$O$_3$, can also be used to form aqueous biphasic systems with hydrophilic ionic liquids.

Such aqueous biphasic systems can potentially be used to salt out and therefore separate hydrophilic ionic liquids in various applications. The ionic liquids can then be recycled.

5.12 MUTUALLY IMMISCIBLE IONIC LIQUIDS

In 2006, Earle and co-workers reported the discovery of room-temperature ionic liquids that are mutually immiscible.[21] The researchers found that mixtures of some hydrophilic and hydrophobic ionic liquids form stable two-phase mixtures, particularly when the structures of the cations in the ionic liquids are dissimilar.

The biphasic systems typically consist of a hydrophobic phosphonium-based ionic liquid such as [P$_{6\ 6\ 6\ 14}$]Cl and a hydrophilic imidazolium ionic liquid such as [C$_2$mim]Cl. The two phases are not pure ionic liquids, however, because some imidazolium cations move from the lower phase into the upper phosphonium phase and some of the phosphonium cations dissolve in the hydrophilic imidazolium layer.

Earle's team also showed that organic compounds, and other solutes such as iodine, greatly prefer one ionic liquid phase over the other.

Mixtures of two hydrophobic ionic liquids with a common anion and cations with significant structural differences, *e.g.* [P$_{6\ 6\ 6\ 14}$][NTf$_2$] and [C$_2$mim][NTf$_2$], also exhibit mutual immiscibility. The systems have an UCST with a phase diagram of the type shown in Figure 5.2.

Some of these biphasic systems are also immiscible with water and alkanes. For example, Earle and co-workers prepared a stable tetraphasic mixture consisting of a top pentane layer, [P$_{6\ 6\ 6\ 14}$][NTf$_2$], water and a bottom layer of [C$_2$mim][NTf$_2$].

The discovery of biphasic ionic liquid systems could potentially be exploited for the separation of organic mixtures, counter-current extraction and applications such as battery technology where a permeable interface is required.

5.13 SEPARATIONS AND EXTRACTIONS

The lack of vapour pressure and the wide variations in miscibility and immiscibility of ionic liquid solvents offer several advantages over traditional organic solvents not only for combined reaction and separation schemes but also for the separation or extraction of materials produced by processes that do not involve ionic liquids.

Biphasic catalysis using an ionic liquid to immobilize a catalyst is an example of a combined reaction and separation scheme. Liquid reaction products that form a separate phase from the ionic liquid–catalyst phase can be simply separated in a laboratory by decantation or the use of a separating funnel. The ionic liquid and catalyst can then be recycled. Hydrophilic reaction products dissolved in a hydrophobic ionic liquid can be removed with water. Thermally-stable volatile organic reaction products can be readily separated from the ionic liquid by distillation without the formation of an azeotrope. Volatile organic compounds can also be removed from ionic liquids by stripping, *i.e.* passing an inert gas or vapour through the ionic liquid to strip out the volatile compounds.

Specific examples of the use of ionic liquids in combined reaction and separation schemes are described in later chapters. The following sections outline some examples of how ionic liquids can be exploited to extract one or more components from a mixture of products produced by a process that does not involve an ionic liquid.

5.14 EXTRACTION OF ORGANIC COMPOUNDS

Unsaturated organic compounds, particularly aromatic compounds, tend to be soluble in ionic liquids whereas saturated organic compounds are generally immiscible with ionic liquids. As a result, it should, in principle, be possible to separate the aromatic and aliphatic hydrocarbons produced by naphtha crackers. Naphtha is a mixture of hydrocarbons obtained by the fractional distillation of crude oil in petroleum refineries. A significant proportion of naphtha is used as feedstock to manufacture other chemicals. Crackers convert the naphtha into mixtures of aromatic hydrocarbons and aliphatic compounds that are then separated by various means.

Meindersma and co-workers have shown that ionic liquids can be employed as alternatives to conventional organic solvents for the liquid–liquid extraction of aromatic hydrocarbons from aromatic-aliphatic mixtures.[22] The group showed that 1-ethyl-3-methylimidazolium ethyl sulfate, $[C_2mim][C_2H_5OSO_3]$, and 1,3-dimethylimidazolium methyl sulfate, $[C_1mim][CH_3OSO_3]$, are suitable for the extraction of toluene from mixtures of toluene and heptane. Toluene and other aromatic hydrocarbons are soluble in these ionic liquids whereas aliphatic hydrocarbons such as alkanes and cyclo-alkanes are not.

Ionic liquids have been employed to extract various organic solutes from aqueous solutions. Wang and colleagues, for example, have shown that the hydrophobic nature of imidazolium ionic liquids with fluorinated anions can be used to extract a range of amino acids from aqueous solutions.[23] They found that ionic liquids with the $[BF_4]^-$ anion exhibit higher extraction efficiency than those with the $[PF_6]^-$ anion. The researchers also showed that increasing the length of the alkyl chain on the cation lowers the level of amino acid extraction.

The pH also plays a crucial role. Wang's team observed higher degrees of extraction of the amino acids from solutions with a low pH. They suggested that this phenomenon arises from the predominance of the cationic forms of the amino acids at low pH, resulting in strong electrostatic interactions between the amino acid cations and the anions of the ionic liquid. Their results indicate that $[C_6mim][BF_4]$ and $[C_8mim][BF_4]$ may be suitable for practical applications. The work also illustrates how ionic liquids can be designed for practical extraction processes.

In another potential application, Dionysiou, Botsaris and co-workers have demonstrated that it is possible to extract organic contaminants, such as chlorophenols, from water using hydrophobic ionic liquids.[24] Chlorophenols are water-soluble compounds that have been used as pesticides, wood preservatives and disinfectants. Natural waters contaminated with chlorophenols pose a health risk to humans and also endanger the biota in the water.

The researchers studied the distribution ratios of chlorophenols in ionic liquid–water mixtures for various ionic liquids. They found that it is technically feasible to use ionic liquids such as [C₄mim][PF₆] and [C₂mim][beti] to extract chlorophenols from water. However, as these ionic liquids are fluorinated compounds and to some extent soluble in water they may not be suitable, from an environmental point of view, for removing chlorinated contaminants from water. The aim now is to find or design environmentally-friendly ionic liquids that do the same job.

In some cases, the extraction of organic solutes from aqueous media can be enhanced by the addition of an organic extractant to the ionic liquid. For example, Matsumoto and co-workers showed that the ability of imidazolium ionic liquids with [PF₆]⁻ anions to extract organic acids, such as lactic acid, is poor.[25] The addition of a conventional extractant like tri-*n*-butylphosphate to the ionic liquid raises the extractability of the organic acids to a level comparable to that of traditional molecular organic solvents.

5.15 GAS SEPARATIONS

The widely variable solubilities of gases in ionic liquids open up the possibility of using ionic liquids as absorption liquids to separate one or more gases from mixtures of gases. The prospect is particularly attractive because ionic liquids are non-volatile and therefore do not evaporate and contaminate the gaseous mixture.

Davis, Jr. and co-workers have designed a task-specific ionic liquid (TSIL) for capturing CO_2 from sour gas.[26] Sour gas is natural gas, CH_4, that is contaminated with relatively high levels of other gases such as CO_2 and hydrogen sulfide, H_2S. When the levels of contaminants are low, the gas is known as sweet gas. The TSIL designed by the group has [BF₄]⁻ as its anion and an imidazolium cation with an amine functional group covalently attached to it. Evidence from nuclear magnetic resonance (NMR) spectroscopy of the TSIL containing the sequestered CO_2 suggests that an ammonium carbamate salt is formed in the sequestering process (Scheme 5.1).

The CO_2-capturing process is reversed by heating under reduced pressure. The recovered ionic liquid can then be recycled for CO_2 uptake without loss of efficiency.

The TSIL is highly viscous, however. Its potential use for large-scale gas scrubbing applications is therefore probably limited. Even so, the team suggested that variants of the TSIL with enhanced chemical and physical properties might be designed for this task.

Ionic liquids can also absorb sulfur dioxide, SO_2. The gas has various uses. For example, it is used as a preservative and in the manufacture of

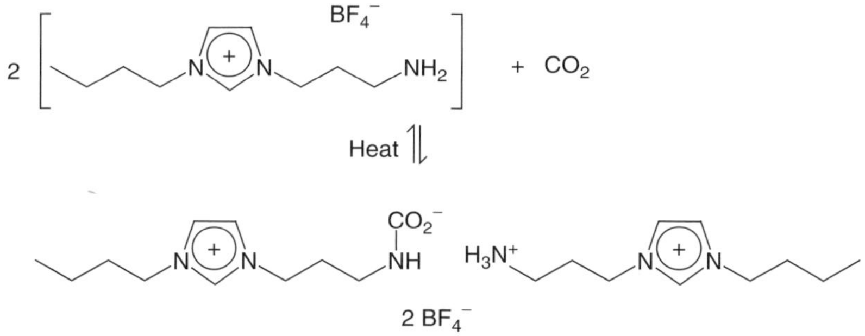

Scheme 5.1 Task-specific ionic liquid sequesters CO_2 to form an ammonium carbamate.

sulfuric acid. However, it is also emitted by volcanoes, by combustion of sulfur-containing fossil fuels, and other industrial processes. When it escapes into the atmosphere, it not only poses a human health problem but also can cause acid rain and smog. Industrial emissions are reduced by processes such as flue gas desulfurization, a process that employs lime (CaO) to react with the gas.

Riisager and co-workers have shown that ionic liquids with 1,1,3,3-tetramethylguanidinium (**5.1**) or $[C_4mim]^+$ cations and $[BF_4]^-$ or $[NTf_2]^-$ anions can rapidly physically absorb large amounts of the SO_2.[27] The gas is readily removed from the ionic liquids by heating at reduced pressure, enabling the ionic liquids to be used again.

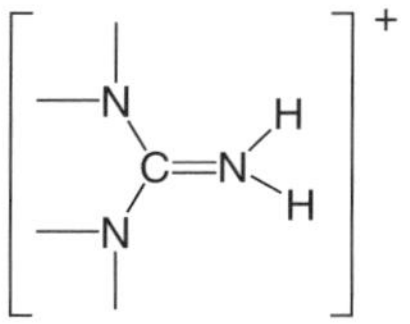

5.1

5.16 CARBON DIOXIDE EXTRACTIONS AND SEPARATIONS

The extraction of non-volatile organic compounds dissolved in ionic liquids without the use of conventional organic solvents is one of the major technical challenges for the use of ionic liquids in chemical manufacturing. Distillation or gas stripping cannot be used to remove non-volatile solutes from ionic liquids.

The use of supercritical CO_2 could possibly overcome this problem. CO_2 dissolves in ionic liquids to form a two-phase system (Figure 5.5). Supercritical CO_2, that is carbon dioxide that has been heated and pressurized above its critical temperature and pressure, is an environmentally-benign solvent. The fluid is not only abundant and inexpensive, it is also non-toxic and non-flammable.

In 1999, Brennecke and co-workers showed that supercritical CO_2 can be used to extract naphthalene, a non-volatile compound, dissolved in $[C_4mim][PF_6]$.[28] The researchers achieved a perfectly clean separation of the two phases. They could find no measurable amount of the ionic liquid in the CO_2 extract.

Two years later, Blanchard and Brennecke reported that supercritical CO_2 can be used to extract quantitatively a range of volatile and non-volatile aromatic organic solutes from $[C_4mim][PF_6]$.[29] In all cases, no appreciable amount of the ionic liquid dissolved in the CO_2 phase.

Brennecke's team has also shown that CO_2 at low pressures can be used to separate a mixture of $[C_4mim][PF_6]$ and methanol,[30] and separate hydrophobic and hydrophilic imidazolium ionic liquids from water.[31] In subsequent work, the team used CO_2 to precipitate inorganic and organometallic compounds from organic–ionic liquid mixtures.[32]

5.17 METAL ION EXTRACTIONS

The use of hydrophobic ionic liquids to replace hydrophobic volatile organic solvents as extracting phases to remove metal ions from aqueous solutions is an exciting prospect. Potential applications include the extraction of metals from ores, the removal of metal contaminants from water, and the reprocessing of nuclear fuels and nuclear waste. However, the affinity of metal ions for hydrophobic ionic liquids is generally low because the ions tend to be hydrated in aqueous solutions.

The problem can be overcome by using conventional metal ion extractants, such as a crown ethers. These extractants dehydrate the metal ions and form complexes with them. The metal-extractant complexes are more hydrophobic than the metal ions and therefore partition preferentially to the hydrophobic solvent. Dicyclohexan-18-crown-6 (DCH18C6) (**5.2**) and other crown ethers have been used to extract alkali metal ions such as Li^+ and Na^+, alkaline earth metal ions such as Mg^{2+}, Ca^{2+} and Sr^{2+}, and heavy metal ions such as Pb^{2+} into ionic liquids

5.2

In one piece of work on this topic, Dietz and co-workers studied the use of DCH18C6 and other extractants to transfer metal ions from acidic aqueous solutions into imidazolium ionic liquids.[33] They concluded that the extraction of the metal ions using such metal-ion extractants into ionic liquids parallels, to some extent, their extraction into conventional solvents.

The nature of the ionic liquid cations and anions influences the extraction efficiency. For example, the solubility of a crown ether–metal complex in ionic liquids can be tuned by altering the length of the alkyl chain on the ionic liquid imidazolium cation. In general, the selectivity for a specific metal ion, *e.g.* the Ba^{2+} ion from a mixture of Ba^{2+}, Sr^{2+}, Ca^{2+} and Mg^{2+} ions in aqueous solution, increases as the alkyl chain length increases. Extraction efficiency, in contrast, decreases with increasing chain length. In addition, the extraction efficiency of crown ether–metal complexes into ionic liquids with the $[NTf_2]^-$ anion is higher than extraction into ionic liquids with the $[PF_6]^-$ anion.

Various other types of conventional metal-ion extractants have also been tested for use with ionic liquids. For example, the metal ion extractants 1-(2-pyridylazo)-2-naphthol (**5.3**) and 1-(2-thiazolylazo)-2-naphthol (**5.4**) can be used to extract transition metal cations, such as Fe^{3+}, Co^{2+} and Cd^{2+}, from basic aqueous solutions to the ionic liquid $[C_6mim][PF_6]$. The metal ions can then be stripped from the ionic liquid phase under acidic conditions.

5.3   **5.4**

Task-specific imidazolium ionic liquids have been designed that simultaneously serve as the extractant and the hydrophobic extracting solvent.[34] A functional group such as thioether **5.5**, urea **5.6** or thiourea **5.7** that acts at the metal-extractant is grafted onto the imidazolium cation. TSILs with these functionalized cations and $[PF_6]^-$ anions are used either alone or mixed with $[C_4mim][PF_6]$ to reduce their viscosity. The TSILs and their mixtures with $[C_4mim][PF_6]$ were shown to be particularly effective at removing Hg^{2+} and Cd^{2+} ions from water.

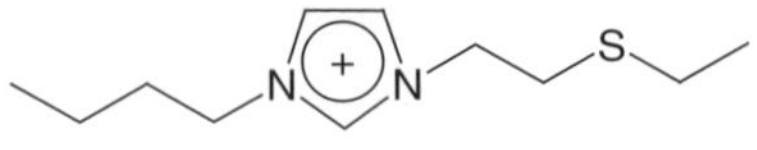

5.5

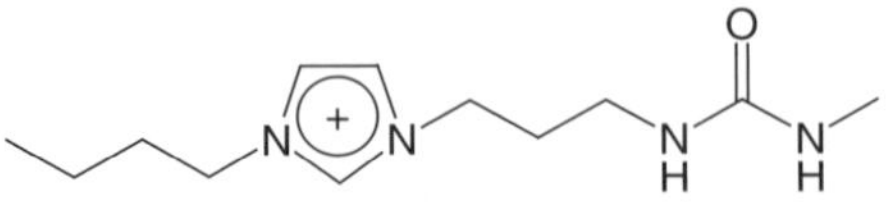

5.6

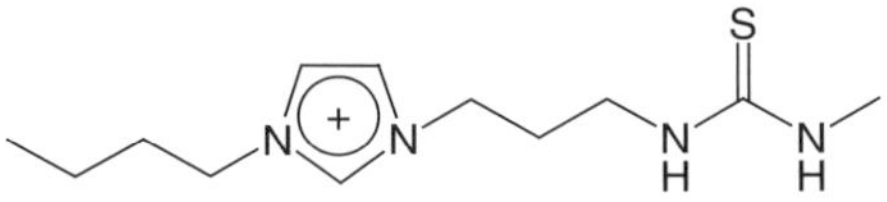

5.7

Metal-extracting TSILs are less soluble in water than conventional hydrophobic ionic liquids. Leaching of the ionic liquids into water is thus minimized. In addition, TSILs are potentially attractive for metal-ion extractions because their functional groups can, in principle, be tailored to extract specific metal ions. The major disadvantage, however, is the difficulty in stripping the metal ions from the TSIL.

5.18   EXTRACTIVE DISTILLATIONS

Ionic liquids can be used as extractants in a liquid–liquid separation process known as extractive distillation. The process uses an entrainer, also known as a mass separating agent or separation enhancer, which enhances the separation of a mixture during fractional distillation. An entrainer increases the difference in volatility between the components of a mixture and is therefore useful in separating the components of azeotropic systems. Azeotropes are mixtures of two or more compounds that cannot be further separated by fractional distillation because the composition of their vapour phase is the same as that of their liquid phase.

Conventional entrainers, such as dimethylformamide, are regenerated after use by distillation. Ionic liquids, on the other hand, have negligible vapour pressures and therefore do not need to be distilled when used as entrainers. Since many ionic liquids have an affinity for water, they are particularly useful as entrainers for the separation of water-containing azeotropes. For example, commercially-available imidazolium ionic liquids, such as [C$_4$mim][BF$_4$], have been shown to be suitable as entrainers for separating the azeotropic ethanol–water system by extractive distillation.[35]

5.19 MEMBRANE SEPARATIONS

The use of porous membranes impregnated with ionic liquids to separate gases has attracted interest over recent years. Supported ionic liquid membranes (SILMs) operate on the same principle as supported liquid membranes (SLMs). A feed stream consisting of a mixture of gases under pressure is passed over one side of the membrane. Soluble gas molecules dissolve in the liquid inside the membrane pores. The molecules then permeate through the membrane by diffusion and desorb on the other side where they are swept out by a receiving gas in the so-called permeate stream.

The major advantage of SILMs over SLMs is the lack of vapour pressure of the liquids. The ionic liquids do not evaporate into the feed or receiving streams whereas the solvents used in conventional SLMs do. The opportunity to design ionic liquids for specific gas separations is an added advantage.

The performance of a gas separation membrane is largely determined by the membrane's transport properties, *i.e.* its permeability and selectivity for a specific gas in a mixture. Permeability is the rate of diffusion of the gas under a pressure gradient through the membrane. Ideally, membranes should exhibit high permeability and high selectivity for a gas. For most membranes, however, there is a trade-off between permeability and selectivity: as permeability increases, selectivity decreases and *vice versa*.

In 2004, Noble and co-workers examined the performance of SILMs consisting of ionic liquids immobilized in porous hydrophilic polyethersulfone membranes in terms of their permeabilities and selectivities for separating CO_2 from CO_2/N_2 and CO_2/CH_4 mixtures.[36] The researchers tested imidazolium and phosphonium ionic liquids with various anions and concluded that their SILMs were effective for separating the mixtures, although the performance depended on the ionic liquid used.

Bara, Noble and colleagues have also shown that ionic liquids consisting of $[NTf_2]^-$ anions and imidazolium cations tailored with a polymerizable group can be converted into solid polymer films that can be used as gas separation membranes.[37] The films performed as well or better than other polymers for the separation of CO_2/N_2 mixtures. Their performance was less impressive for separating CO_2/CH_4 mixtures.

Gas separation membranes typically have micrometer-size pores. In 2006, Gan and co-workers described an investigation into the gas transport and separation properties of ionic liquids supported on nanofiltration membranes.[38] These membranes have smaller pore sizes and are generally made of speciality polymers such as polyimides. They are used for the separation of low molecular weight solutes and small inorganic ions.

The investigation revealed that SILMs based on a commercially available nanofiltration membrane and four ionic liquids, [C$_4$mim] [NTf$_2$], [C$_{10}$mim][NTf$_2$], [N$_8$ $_8$ $_8$ $_1$][NTf$_2$] and [C$_8$Py][NTf$_2$], are "remarkably stable" even at high pressure. The researchers investigated the permeation rate of H$_2$, N$_2$, O$_2$ and CO in the SILMs and also the selectivity of H$_2$ *versus* CO for a mixture of the two gases. All four ionic liquids exhibited a degree of H$_2$/CO selectivity although the value varied, depending on the nature of the cation. The permeation rate depended on the viscosity and molecular weight of the ionic liquids. [N$_8$ $_8$ $_8$ $_1$][NTf$_2$] had the best H$_2$/CO selectivity but its permeation rate was far less than that of [C$_8$Py][NTf$_2$], which had the best permeation performance but the worst selectivity.

SILMs can also be used to separate liquids. For example, ionic liquids with [C$_4$mim]$^+$, [C$_8$mim]$^+$ or [C$_{10}$mim]$^+$ cations and [BF$_4$]$^-$ or [PF$_6$]$^-$ anions immobilized on hydrophilic polymeric membranes have been shown to achieve good selectivity for a specific organic solute.[39] In particular, [C$_4$mim][PF$_6$] immobilized in a poly(vinylidene fluoride) (PVDF) membrane had an extremely high selectivity ratio of 55 : 1 for secondary amines over tertiary amines. SILMS consisting of imidazolium or tetraalkylammonium ionic liquids immobilized in a PVDF membrane can also transport selectively aromatic hydrocarbons such as benzene, toluene and *p*-xylene.[40]

REFERENCES

1. Y. Patell, K. R. Seddon, L. Dutta and A. Fleet, in *Green Industrial Applications of Ionic Liquids*, ed. R. D. Rogers, K. R. Seddon and S. Volkov, NATO Science Series, Kluwer Academic Publishers, Dordrecht, 2002, p. 499.
2. C. Reichardt, *Green Chem.*, 2005, **7**, 339.
3. L. A. Blanchard, Z. Gu, J. F. Brennecke and E. J. Beckman, in *Green Industrial Applications of Ionic Liquids*, ed. R. D. Rogers, K. R. Seddon and S. Volkov, NATO Science Series, Kluwer Academic Publishers, Dordrecht, 2002, p. 403.
4. P. Wasserscheid and W. Keim, *Angew. Chem. Int. Ed.*, 2000, **39**, 3772.
5. J. D. Holbrey, W. M. Reichert, M. Nieuwenhuyzen, O. Sheppard, C. Hardacre and R. D. Rogers, *Chem. Commun.*, 2003, 476.

6. C. H. Hussey, *Pure & Appl. Chem.*, 1988, **60**, 1763.
7. J. L. Anthony, E. J. Maginn and J. F. Brennecke, *J Phys. Chem. B*, 2001, **105**, 10942.
8. J. L. Anthony, E. J. Maginn and J. F. Brennecke, *J. Phys. Chem. B*, 2002, **106**, 7315.
9. J. Jacquemin, M. F. Costa Gomes, P. Husson and V. Majer, *J. Chem. Thermodyn.*, 2006, **38**, 490.
10. J. Kumelan, Á. P.-S. Kamps, D. Tuma and G. Maurer, *J. Chem. Eng. Data*, 2006, **51**, 1364.
11. R. E. Baltus, B. H. Culbertson, S. Dai, H. Luo and D. W. DePaoli, *J. Phys. Chem. B*, 2004, **108**, 721.
12. C. Cadena, J. L. Anthony, J. K. Shah, T. I. Morrow, J. F. Brennecke and E. J. Maginn, *J. Am. Chem. Soc.*, 2004, **126**, 5300.
13. J. L. Anderson, J. L. Anthony, J. F. Brennecke and E. J. Maginn, in *Ionic Liquids in Synthesis*, 2nd edn, ed. P. Wasserscheid and T. Welton, Wiley-VCH, Weinheim, Germany, 2008, **vol. 1**, p. 103.
14. S. N. V. K. Aki, D. G. Hert, J. L. Anderson and J. F. Brennecke, 1st International Congress on Ionic Liquids, Salzburg, Austria, June 19–22, 2005, Book of Abstracts, p. 104.
15. A. A. Fannin Jr., D. A. Floreani, L. A. King, J. S. Landers, B. J. Piersma, D. J. Stech, R. L. Vaughn, J. S. Wilkes and J. L. Williams, *J. Phys. Chem.*, 1984, **88**, 2614.
16. J. Lachwa, J. Szydlowski, V. Najdanovic-Visak, L. P. N. Rebelo, K. R. Seddon, M. Nunes da Ponte, J. M. S. S. Esperança and H. J. R. Guedes, *J. Am. Chem. Soc.*, 2005, **127**, 6542.
17. J. E. L. Dullius, P. A. Z. Suarez, S. Einloft, R. F. de Souza and J. Dupont, *Organometallics*, 1998, **17**, 815.
18. J. G. Huddleston, H. D. Willauer, R. P. Swatloski, A. E. Visser and R. D. Rogers, *Chem. Commun.*, 1998, 1765.
19. A. E. Visser, R. P. Swatloski and R. D. Rogers, *Green Chem.*, 2000, **2**, 1.
20. K. E. Gutowski, G. A. Broker, H. D. Willauer, J. G. Huddleston, R. P. Swatloski, J. D. Holbrey and R. D. Rogers, *J. Am. Chem. Soc.*, 2003, **125**, 6632.
21. A. Arce, M. J. Earle, S. P. Katdare, H. Rodriguez and K. R. Seddon, *Chem. Commun.*, 2006, 2548.
22. G. W. Meindersma, A. J. G. Podt, M. F. Meseguer and A. B. de Haan, in *Ionic Liquids IIIB: Fundamentals, Progress, Challenges, and Opportunities. Transformations and Processes*, ed. R. D. Rogers and K. R. Seddon, ACS Symposium Series 902, American Chemical Society, Washington DC, 2005, p. 57.
23. J. Wang, Y. Pei, Y. Zhao and Z. Hu, *Green Chem.*, 2005, **7**, 196.

24. E. Bekou, D. D. Dionysiou, R.-Y. Qian and G. D. Botsaris, in *Ionic Liquids as Green Solvents: Progress and Prospects*, ed. R. D. Rogers and K. R. Seddon, ACS Symposium Series 856, American Chemical Society, Washington DC, 2003, p. 544.

25. M. Matsumoto, K. Mochiduki, K. Fukunishi and K. Kondo, *Sep. Purif. Technol.*, 2004, **40**, 97.

26. E. D. Bates, R. D. Mayton, I. Ntai and J. H. Davis Jr., *J. Am. Chem. Soc.*, 2002, **124**, 926.

27. J. Huang, A. Riisager, P. Wasserscheid and R. Fehrmann, *Chem. Commun.*, 2006, 4027.

28. L. A. Blanchard, D. Hancu, E. J. Beckman and J. F. Brennecke, *Nature*, 1999, **399**, 28.

29. L. A. Blanchard and J. F. Brennecke, *Ind. Eng. Chem. Res.*, 2001, **40**, 287.

30. A. M. Scurto, S. N. V. K. Aki and J. F. Brennecke, *J. Am. Chem. Soc.*, 2002, **124**, 10276.

31. A. M. Scurto, S. N. V. K. Aki and J. F. Brennecke, *Chem. Commun.*, 2003, 572.

32. B. R. Mellein, E. M. Saurer, S. N. V. K. Aki and J. F. Brennecke, 1st International Congress on Ionic Liquids, Salzburg, Austria, June 19–22, 2005, Book of Abstracts, p. 64.

33. M. L. Dietz, J. A. Dzielawa, M. P. Jensen, J. V. Beitz and M. Borkowski, in *Ionic Liquids IIIB: Fundamentals, Progress, Challenges, and Opportunities. Transformations and Processes*, ed. R. D. Rogers and K. R. Seddon, ACS Symposium Series 902, American Chemical Society, Washington DC, 2005, p. 2.

34. A. E. Visser, R. P. Swatloski, W. M. Reichert, R. Mayton, S. Sheff, A. Wierzbicki, J. H. Davis, Jr. and R. D. Rogers, *Environ. Sci. Technol.*, 2002, **36**, 2523.

35. C. Jork, M. Seiler, Y.-A. Beste and W. Arlt, *J. Chem. Eng. Data*, 2004, **49**, 852.

36. P. Scovazzo, J. Kieft, D. A. Finan, C. Koval, D. DuBois and R. Noble, *J. Membr. Sci.*, 2004, **238**, 57.

37. J. E. Bara, S. Lessmann, C. J. Gabriel, E. S. Hatakeyama, R. D. Noble and D. L. Gin, *Ind. Eng. Chem. Res.*, 2007, **46**, 5397.

38. Q. Gan, D. Rooney, M. Xue, G. Thompson and Y. Zou, *J. Membr. Sci.*, 2006, **280**, 948.

39. L. C. Branco, J. G. Crespo and C. A. M. Afonso, *Chem.–Eur. J.*, 2002, **8**, 3865.

40. M. Matsumoto, Y. Inomoto and K. Kondo, *J Membr. Sci.*, 2005, **246**, 77.

CHAPTER 6

Green Credentials of Ionic Liquids

6.1 INTRODUCTION

Chemistry, more than any other branch of science, "has provided a cornucopia of good things, both of necessities and luxuries, which have improved our health, our wealth, and also I believe our happiness" said Lord George Porter (1920–2002) in a plenary lecture at the 30th Congress of the International Union of Pure & Applied Chemistry (IUPAC) held in Manchester, UK, in 1985.[1] Porter, who won the Nobel Prize for Chemistry in 1967, gave numerous examples of chemistry's cornucopia. He pointed out, for example, that during the twentieth century our health improved out of all recognition and our lifespan doubled through better nutrition, better hygiene and the use of an ever-increasing array of pharmaceutical products.

Yet chemistry's cornucopia, which we all enjoy, has several downsides. They include pollution, depletion of non-renewable resources and the wastage of materials. Chemistry and technology that aim to minimize these downsides are known as green chemistry and clean technology.

The prospect of using room-temperature ionic liquids for clean technology was recognized by Seddon in the mid-1990s. In a paper published in 1996 on the use of chloroaluminate ionic liquids for clean catalysis, he wrote: "Clean technology is concerned with reducing the waste from an industrial chemical process to zero. Its implementation will lead to a cleaner environment and more cost-effective use of starting materials."[2] He suggested that ionic liquids are neoteric (ground

An Introduction to Ionic Liquids
By Michael Freemantle
© Michael Freemantle 2010
Published by the Royal Society of Chemistry, www.rsc.org

breaking) solvents that have several advantages over the volatile organic molecular solvents traditionally used in industrial chemical processes and alternative solvents such as water. He listed seven "remarkable" new properties of ionic liquids that are "desirable" for clean technology:

- a liquid range of 300 °C compared with 100 °C for water;
- they are good solvents for a wide range of inorganic, organic and polymer materials;
- they exhibit Brønsted, Lewis and other types of acidity;
- they have no effective vapour pressure;
- their water sensitivity does not restrict their industrial applications;
- thermal stability up to 200 °C;
- they are relatively inexpensive and easy to prepare.

Seddon elaborated the design principles for using chloroaluminate ionic liquids as solvents for clean technology in another paper the following year.[3] He observed that the cation or anion of an ionic liquid can be changed, if not at will then certainly with considerable ease, so as to optimize the relative solubilities of the reactants and products, its liquid range, the reaction kinetics, the catalytic behaviour of the media and the cost. In both papers, he used the term "designer solvents" for ionic liquids.

In 1998, an article in *Chem & Eng. News* drew the attention of the concept of ionic liquids as "designer solvents" to a much wider audience.[4] In that article, Anastas, who, at the time, was chief of the industrial chemistry branch at the Environmental Protection Agency (EPA), in Washington D.C., commented that he was convinced that ionic liquids could contribute significantly to green chemistry and the development of clean technology.

In the same year, Anastas and Warner stated 12 principles of green chemistry in a book on the theory and practice of green chemistry.[5]

The book has a brief reference to ionic liquids but in the section "Examples of green solvents and reaction conditions" there are no examples of the use of ionic liquids for green chemistry, although there are several examples of the use of supercritical fluids and aqueous media. At that time, little research had been carried out on the development of ionic liquids specifically for green chemistry and clean technology.

The design of chemical processes that use safer eco-friendly solvents, such as ionic liquids or water, is one of the principles upon which green chemistry is founded. Others include:

- designing processes that produce useful chemicals without producing waste;

- designing chemicals and chemical products that have little or no toxicity;
- using renewable raw materials and feedstock for chemical processes;
- where possible, increasing energy efficiency by carrying out chemical processes at ambient temperature and pressure;
- designing chemicals and products to degrade into benign substances that do not accumulate in the environment;
- designing chemicals and chemical processes that minimize the risk of fires, explosions and other accidents.

In 2000, Earle and Seddon published a paper in the IUPAC journal *Pure & Applied Chemistry*, stating that the lack of measurable vapour pressure of ionic liquids "characterizes them as green solvents".[6] They noted that choice of the correct ionic liquid as a solvent can result in high product yields and reduced amounts of waste. In addition, the ionic liquid can be recycled, leading to reduced costs of processes. Reactions are also often quicker and easier to carry out in ionic liquids than in conventional organic solvents.

The term "green solvent" was subsequently applied to ionic liquids in several publications. For example, Eckert and co-workers published a paper entitled: "Ionic liquids as catalytic green solvents for nucleophilic displacement reactions."[7] They showed that the reaction of potassium cyanide with benzyl chloride to yield phenylacetonitrile could be carried out in the "environmentally benign solvent" [C_4mim][PF_6]. The process eliminates the need for a volatile organic solvent and hazardous catalyst disposal, the authors observed.

Chemists soon came to realise, however, that the concept that all ionic liquids are "green solvents" is misleading, if not erroneous. Many ionic liquids have attractive "green" properties but they can also have "non-green" properties, most notably toxicity. In addition, little is known about the potential impact of ionic liquids on the environment when they are discharged as waste or when a spillage occurs.

Furthermore, many of the most widely used ionic liquids contain imidazolium-based cations. These ionic liquids are typically synthesized from imidazoles and halogenoalkanes (Chapter 3), both of which derive from non-renewable and "non-green" petroleum feedstocks. In addition, organic solvents are usually needed for their synthesis.

Anions such as [PF_6]$^-$ are synthesized using toxic salts and energy-intensive electrochemical processes. However, several ionic liquids are known that can be synthesized from renewable resources. For example, ionic liquids with cations based on nicotinic acid have been synthesized.[8]

Nicotinic acid (pyridine-3-carboxylic acid, $C_5H_4NCO_2H$) is also known as niacin or vitamin B_3. The compound occurs naturally in various foods.

Nowadays, it is widely recognized that ionic liquids are not intrinsically green. Even so, there is general acceptance that they can contribute significantly to the development of green chemistry. They can become part of the green chemistry tool box. For example, ionic liquids can, in many cases, increase the rates of reactions compared with conventional molecular solvents. As a result, energy input is reduced and energy efficiency increased. In addition, in many reactions, ionic liquids are less wasteful than conventional solvents because they give higher yields and are more selective for a desired product. Some examples of these reactions are outlined in later chapters.

Ionic liquids can also prevent or reduce waste in other ways. For example, aromatic nitrations in industry employ concentrated nitric acid as a nitrating agent and concentrated sulfuric acid as a catalyst (Section 10.16). Large amounts of environmentally harmful acidic and water-reactive wastes are generated in the process. In ionic liquids, however, the nitrations can be carried out using milder and more environmentally benign nitrating agents and without the production of aqueous acidic waste.

The extent to which ionic liquids can be exploited for green chemistry and clean technology will depend on the choice or design of ionic liquids for specific applications and also on the acquisition of data on the chemical, physical and biological properties of the ionic liquids. However, compared with other families of chemical substances, the study of ionic liquids is in its infancy. There are little data relating to the green credentials of many ionic liquids – such as toxicity, biological oxygen demand and biodegradability. Even so, scientific knowledge about ionic liquids and their possible contribution to green chemistry and clean technology is rapidly developing. Even one of the greenest properties of ionic liquids, if not the greenest, *i.e.* negligible vapour pressure, has come under scrutiny in recent years.

6.2 VAPOUR PRESSURE

Ionic liquids do not exert measurable vapour pressures at ambient temperatures. Because they do not emit vapours at these temperatures, they cannot pollute the atmosphere. They are therefore perceived as potential replacements for volatile organic compounds. As a result, negligible vapour pressure has become the hallmark property that singles out ionic liquids for the development of green chemistry.

For many years, it was widely assumed that this general property, the lack of measurable vapour pressure, meant that ionic liquids could not be distilled. In 2006, however, an international team of researchers led by Rebelo demonstrated that this assumption is unfounded.[9]

The researchers showed that a range of pure aprotic dialkylimidazolium, tetraalkylammonium and tetraalkylphosphonium ionic liquids can be evaporated under vacuum at 200–300 °C without thermal decomposition and then recondensed at lower temperatures. They also demonstrated that equimolar binary mixtures of the ionic liquids can be distilled. In all cases, the composition of the distillate from the mixtures was enriched in one component and the composition of the residue enriched in the other component.

The discovery shows that distillation might, in principle at least, be used to purify ionic liquids and also regenerate used ionic liquids. It also shows that ionic liquids could potentially participate in gas-phase processes.

Many ionic liquids, however, do not distil below their thermal decomposition temperatures. Their negligible vapour pressures at lower temperatures enables volatile organic reaction products to be extracted by distillation, allowing the ionic liquids to be recycled.

Their lack of vapour pressure not only makes ionic liquids attractive as potential replacements for volatile organic solvents. They can also be used to circumvent the volatility of toxic reagents and by-products. The Swern oxidation of alcohols (Section 11.14) provides an example. The oxidation normally involves the use of dimethyl sulfoxide (DMSO) as the oxidant and generates dimethyl sulfide (DMF) as a by-product. DMSO and DMF are both volatile and toxic. Ionic liquids side-step this problem by anchoring the sulfoxide oxidant and the sulfide, thereby rendering them involatile.

Ionic liquids can also be used to store and deliver hazardous gases such as phosphine safely under ambient conditions (Section 14.5).

Although ionic liquids do not exert measurable vapour pressure at room temperature and pressure, this does not mean that all ionic liquids can be safely handled in an open beaker in a laboratory. Haloaluminate ionic liquids, for example, decompose in the presence of water or moisture to form corrosive hydrogen halide fumes. They therefore have to be handled in a dry-box.

Furthermore, it is well-established that ionic liquids with fluorinated anions, such as $[PF_6]^-$, that were originally thought to be air and water stable are unstable towards water and form toxic and corrosive decomposition products such as hydrogen fluoride along with a colourless solid.

One piece of research on this topic has been described by Swatloski and co-workers in a paper entitled: "Ionic liquids are not always green: hydrolysis of 1-butyl-3-methylimidazolium hexafluorophosphate."[10] They characterized the solid as 1-butyl-3-methylimidazolium fluoride hydrate, [C$_4$mim]F.H$_2$O. The authors concluded that ionic liquids, particularly when searching for "green solvents", should be treated as any other research chemical with potentially hazardous properties, unknown toxicity and/or stability.

6.3 FLAMMABILITY AND COMBUSTIBILITY

Unlike organic solvents, ionic liquids are generally considered to be non-flammable and non-combustible and therefore safe to use near flames and heat sources. Flammable liquids are defined as liquids that have flash points below 37.8 °C whereas combustible liquids have flash points above this temperature.[11] The flash point of a liquid is the lowest temperature at which the vapour produced by the liquid will ignite when exposed to an ignition source.

In 2003, Fox and co-workers described an investigation into the flammability, thermal stability and melting points of a range of room-temperature trialkylimidazolium ionic liquids such as 1,2-dimethyl-3-propylimidazolium hexafluorophosphate and 1-butyl-2,3-dimethylimidazolium chloride.[12] They found that the trialkylimidazolium ionic liquids are more thermally stable than the corresponding dialkylimidazolium ionic liquids. The researchers were unable to detect any flash points for any of the trialkylimidiazolium ionic liquids below 200 °C – the maximum temperature of the flashpoint apparatus they used. However, their studies of the correlation between decomposition temperatures and flashpoints for common organic solvents suggested that the flashpoints for imidazolium-based room-temperature ionic liquids with nucleophilic halide anions are "likely to be" in the region of 250–300 °C and around 450 °C for those with larger fluoride-containing anions.

In 2006, Rogers and co-workers published a study of the combustibility of ionic liquids with imidazolium cations and nitrate, picrate or azolate anions.[11] All the ionic liquids tested could be ignited with a small flame torch. The rate of combustion appeared to depend on the nitrogen and oxygen content of the liquids. Some of the ionic liquids burned for a short time and then went out while others burned rapidly to complete or almost complete combustion. The authors concluded that the thermal decomposition products of some ionic liquids are sensitive to combustion. They added that ionic liquids in general should not be assumed to be safe in the proximity of heat or ignition sources.

The combustibility of ionic liquids with oxidizing anions, such as nitrate, and cations that act as a fuel can potentially be exploited for the development of new types of high-energy density materials such as liquid propellants (Section 14.8). The dual nature of ionic liquids should, in principle, enable safe transportation of the propellants. Two ionic liquids, one containing the fuel cations and the other containing the oxidizing anions, could be stored in separate tanks. The propellant is then prepared immediately prior to ignition by mixing the two liquids together.

6.4 RECYCLING

One of the environmentally-beneficial aspects of using ionic liquids as solvents is the possibility of recycling not just the ionic liquids themselves but also catalysts immobilized in the ionic liquids. Recycling not only reduces waste of expensive materials such as transition metal catalysts and the ionic liquids themselves, it also reduces costs.

In a typical scheme, the catalyst is dissolved or immobilized in the ionic liquid. The organic reactants and products form a separate phase or layer. At the end of the reaction, the products are separated by simple decantation. The ionic liquid/catalyst mixture is then re-used. Several examples of such systems are described in later chapters.

Such biphasic catalytic systems can also lead to energy savings compared with homogeneous catalysis in organic solvents where energy intensive processes such as distillation are required to recover the solvents.

However, the use of ionic liquids for biphasic catalysis is not without problems. For example, there are many examples of toxic transition metal catalysts leaching from ionic liquids into products that then must be purified. In addition, impurities can be carried over from cycle to cycle and accumulate in the ionic liquid/catalyst mixture. The catalyst may then become steadily deactivated.

An early example of the recycling power of ionic liquids was described by Andersen and co-workers in 1998.[13] In a paper entitled "Clean catalysis with ionic solvents – phosphonium tosylates for hydroformylation" they described the synthesis of four "high-melting" phosphonium tosylates and their application for the rhodium-catalysed hydroformylation of hex-1-ene. The four ionic compounds have melting points of around 100 °C and are therefore solids at room temperature.

Hydroformylation is the reaction of carbon monoxide and hydrogen with an alkene to form an aldehyde.

The group carried out the reactions in the ionic liquid solvents at 120 °C, achieving high conversions of the hex-1-ene into mixtures of aldehydes. The compositions of the mixtures depended on the substituents attached to the central phosphorus atom of the ionic solvent's cation. After the reactions, the reaction mixtures were simply cooled and the liquid organic products separated by decantation, leaving a solid mixture of the ionic solvent and catalyst. Analysis of the products revealed negligible or no leaching of the rhodium catalyst into the products. The ionic solvent/catalyst system was re-used several times, giving reproducible results.

6.5   ENVIRONMENTAL EXPOSURE AND TOXICITY

Investigations into the safety, health and environment issues soon followed in the wake of the burgeoning interest and usage of ionic liquids in the early 2000s. In particular, researchers realised that information on the toxicity, ecotoxicity and biodegradability of this new class of solvents would be needed to assess their credentials for green chemistry and clean technology.

One of the first papers that focused on environmental exposure appeared in 2003. Jastorff, Ranke and colleagues described a "multi-dimensional risk analysis" of two ionic liquids with imidazolium cations: [C_4mim][BF_4] and [C_{10}mim][BF_4].[14] In 2007, the group followed up their initial work with a comprehensive review of the ecotoxicological risk profiles of numerous ionic liquids.[15]

Their analyses were based on several ecotoxicological indicators, which are given in the following subsections.

6.5.1   Release into the Environment

The risk of release of ionic liquids into the air is low because of their negligible vapour pressures. However, the risk of release into waste-waters and, by accident, directly to soils, surface water or ground water is higher. The risk of release of decomposition and biodegradation products, which may be volatile, and the release of metabolic transformation products into the environment also has to be considered. The risk of release depends to a large extent on the technical systems that employ the ionic liquids and on the control of these systems.

6.5.2 Spatiotemporal Range

The second risk indicator is the tendency of an ionic liquid released into the environment and its environmental transformation products to spread through the environment in space and time. Jastorff and colleagues noted in their 2007 review that data for many ionic liquids is either lacking or not available. They added that the cations and anions will generally have separate fates in the environment and will therefore have to be evaluated separately.

6.5.3 Biodegradability

In 2002, Gathergood and Scammells described an investigation into the biodegradability of ionic liquids with imidazolium cations.[16] They showed that biodegradability increases when an ester group is integrated into the imidazolium side chain. However, the incorporation of the *N*-butyl-*N*-methyl amide group into the side chain did not influence the biodegradability of the ionic liquid.

Three years later, Docherty and co-workers reported a study of the biodegradability of six imidazolium and pyridinium ionic liquids.[17,18] They assessed the ability of microbial communities from two wastewater treatment plants to biodegrade these ionic liquids. Over four weeks, ionic liquids with cations containing octyl substituents exhibited several indicators of biodegradability. Those with butyl or hexyl substituents, although less toxic, were not readily biodegradable.

More recently, Scammells and co-workers carried out a systematic investigation of the biodegradability of pyridinium-based ionic liquids and salts derived from pyridine and nicotinic acid.[8] They reported that 1-alkylpyridinium ionic liquids with linear alkyl chains generally exhibited relatively low levels of biodegradability. However, those with ester-containing substituents at the 1 or 3 positions on the pyridine ring exhibited "excellent" biodegradability.

Jastorff and co-authors in their 2007 review remarked that the biodegradation of the cations and anions should be evaluated separately. The authors observed that, whereas there is a "good amount of data" on the degradability of ionic liquid cations, the lack of knowledge about the fate of widely-used organic anions, such as $[NTf_2]^-$, is "disturbing".

6.5.4 Biological Activity

In their 2003 paper, Jastorff and colleagues reported work on *in vitro* assays of alkylimidazolium ionic liquids using luminescent marine bacteria and mammalian tumour cell lines. They showed that toxicity increases uniformly with increasing alkyl chain length. This is known as the "chain length effect."

The toxicological effects of ionic liquids, notably those with 1-alkyl-3-imidazolium cations, have since been assessed for various enzymes, microorganisms and cell cultures. Jastorff's group, for example, collected data on the cytotoxicities of 253 compounds in a leukaemia rat cell line and values for the inhibition of the enzyme acetylcholinesterase by 292 compounds.[15] The compounds studied covered a wide variety of ionic liquids and closely related salts. From this and other studies, the group concluded that the toxicities of ionic liquids towards cell cultures and microorganisms span the whole range of biocidal potencies, ranging from biocompatibility at very high aqueous concentrations to high biocidal activity.

The toxicity of ionic liquids to various aquatic organisms has also been investigated in recent years. For example, Docherty and co-workers carried out toxicity bioassays to test for imidazolium and pyridinium ionic liquid effects on water fleas, duckweed, snails, minnows, algae and bacteria.[17,18] All the organisms studied exhibited increased ionic liquid toxicity with increasing chain length of alkyls substituted onto the imidazolium and pyridinium cations. Ionic liquids with octyl chains on the cation were more toxic than many volatile solvents currently in use. However, ionic liquids with butyl or hexyl chains showed similar, or lower, toxicity to the volatile solvents. Varying the anion did not significantly alter ionic liquid toxicity.

In another study, Pretti and co-workers determined the acute toxicities of various types of ionic liquid to the zebrafish, *Danio rerio*.[19] The acute toxicities of the ionic liquids were expressed in terms of lethal concentration (LC_{50}) values. These values are the concentrations of the ionic liquids in water that kill 50% of the fish over continuous exposure for 96 hours.

The research showed that imidazolium, pyridinium and pyrrolidinium ionic liquids had a toxic effect on the fish, even at low concentrations, whereas LC_{50} values for ammonium ionic liquids were "remarkably lower" than those reported for conventional organic solvents.

There have been a few of studies of the toxicity of 1-alkyl-3-methylimidazolium ionic liquids to plants and mammals. They show that even low concentrations of these ionic liquids can inhibit plant growth and exhibit acute toxicity to rats, rabbits and mice.

Once again, the chain length effect appears to apply. For example, at the 1st International Congress on Ionic Liquids, Salzburg, Austria, held in 2005, Maase presented a complete toxicological profile of two widely used ionic liquids: 1-ethyl-3-methylimidazolium ethyl sulfate, [C$_2$mim][C$_2$H$_5$SO$_4$], and 1-butyl-3-methylimidazolium chloride, [C$_4$mim]Cl.[18] He reported that [C$_2$mim][C$_2$H$_5$SO$_4$] is not harmful with respect to acute oral toxicity, is non-irritant to skin and eyes, and is non-sensitizing and non-mutagenic. [C$_4$mim]Cl, however, exhibits acute oral toxicity. Neither liquid is readily biodegradable.

Ionic liquids with non-toxic anions, such as lactate, have also been prepared.[20] In 2004, Hoffman, Davis, Jr. and co-workers described the synthesis of a family of ionic liquids consisting of organic cations and the anions of saccharin (**6.1**) or acesulfame (**6.2**) – two non-nutritive food sweeteners.[21] All the salts melt below 100 °C and seven are liquid at room temperature. The sweetener anions are non-toxic and are also not as strongly coordinating towards metals as non-toxic organoanions such as acetate. Furthermore, they are not as basic and do not engage as readily as carboxylates in hydrogen bonding. Two of the ionic liquids synthesized by the group also have the non-toxic cation [(CH$_3$)$_3$N(CH$_2$)$_2$OC(O)(CH$_2$)$_2$CH$_3$]$^+$, formed from butyryl choline.

6.1 **6.2**

In general, however, for many if not most of the ionic liquids that have been reported in the scientific literature there is little or no eco-toxicological data or toxicological information on their possible impact on human health.

6.5.5 Bioaccumulation

The potential bioaccumulation of ionic liquids, particularly toxic ionic liquids, in the environment is an important consideration for the industrial development of ionic liquids.

Jastorff and colleagues noted in their 2003 paper that the tendency of the widely used alkylimidazolium ionic liquids to bioaccumulate is unknown. The group added, however, that those with longer alkyl chain lengths, such as [C$_{10}$mim][BF$_4$], "can be presumed" to have a tendency to be incorporated into membranes. These ionic liquids have a charged head group, the imidazolium moiety, and a nonpolar tail, the alkyl chain. They are therefore structurally similar to membrane lipids. Even so, the limited solubility of these ionic liquids in water should inhibit bioaccumulation. In their 2007 review, the same group noted that the bioaccumulation of typical ionic liquid cations and anions "has yet to be fully investigated."[15]

One measure of potential bioaccumulation of a compound is its partitioning between octanol and water. Brennecke and colleagues have examined the partitioning of [C$_4$mim][NTf$_2$] in this system.[22] They showed that the ionic liquid prefers water over octanol, even though it is immiscible with water at room temperature. Its bioaccumulation potential is therefore quite small.

Stepnowski has also described a study of the potential environmental impact of imidazolium ionic liquids.[23] The study of the sorption of ionic liquids by solids showed that the ionic liquids are strongly and irreversibly bound to most soil types, thus limiting potential contamination of ground water. The lipophilicity of these ionic liquids indicates that they can interact with biological and environmental barriers and that the principal mechanism of toxicity is by disruption of the biological membrane. The disruption is more extensive when the imidazolium cation has a longer alkyl chain. Environmental factors, such as salinity and pH, also have an impact on the toxicity and distribution of the ionic liquids.

Two main conclusions can be drawn concerning the impact of ionic liquids on human health and on the environment. First, there is a dearth of toxicological and ecotoxicological data and information on many ionic liquids. Second, from the data that has been acquired to date, properties such as toxicity and biodegradability vary immensely from ionic liquid to ionic liquid. Even minor structural differences, such as the length of an alkyl side chain, can have a pronounced positive or negative impact on the toxicity of an ionic liquid.

Even so, the large number of existing and potential ionic liquids provide ample opportunity for the design and development of environmentally benign ionic liquids with the requisite physical, chemical and biological properties for specific applications.

Furthermore, properties such as toxicity can be put to good use. For example, the antimicrobial activity of ionic liquids can potentially be exploited for the development of new and improved antiseptics, disinfectants and anti-fouling agents (Section 14.12). The possibility of exploiting the toxicity and dual cation–anion nature of ionic liquids to develop ionic liquid cocktails of active pharmaceutical ingredients customized to the specific needs of patients is also attracting the attention of researchers.

REFERENCES

1. M. Freemantle, *Chemistry in Action*, Macmillan Education Ltd, Basingstoke, 1987, p. 247.
2. K. R. Seddon, *Kinetics Catal.*, 1996, **37**, 693.
3. K. R. Seddon, *J. Chem. Technol. Biotechnol.*, 1997, **68**, 351.
4. M. Freemantle, *Chem. Eng. News*, March 30, 1998, 32.
5. P. T. Anastas and J. C. Warner, *Green Chemistry: Theory and Practice*, Oxford University Press, London, 1998.
6. M. J. Earle and K. R. Seddon, *Pure. Appl. Chem.*, 2000, **72**, 1391.
7. C. Wheeler, K. N. West, C. L. Liotta and C. A. Eckert, *Chem. Commun.*, 2001, 887.
8. J. R. Harjani, R. D. Singer, M. T. Garcia and P. J. Scammells, *Green Chem.*, 2009, **11**, 83.
9. M. J. Earle, J. M. S. S. Esperança, M. A. Gilea, J. N. C. Lopes, L. P. N. Rebelo, J. W. Magee, K. R. Seddon and J. A. Widegren, *Nature*, 2006, **439**, 831.
10. R. P. Swatloski, J. D. Holbrey and R. D. Rogers, *Green Chem.*, 2003, **5**, 361.
11. M. Smiglak, W. M. Reichert, J. D. Holbrey, J. S. Wilkes, L. Sun, J. S. Thrasher, K. Kirichenko, S. Singh, A. R. Katritzky and R. D. Rogers, *Chem. Commun.*, 2006, 2554.
12. D. M. Fox, W. H. Awad, J. W. Gilman, P. H. Maupin, H. C. De Long and P. C. Trulove, *Green Chem.*, 2003, **5**, 724.
13. N. Karodia, S. Guise, C. Newlands and J.-A. Andersen, *Chem. Commun.*, 1998, 2341.
14. B. Jastorff, R. Störmann, J. Ranke, K. Mölter, F. Stock, B. Oberheitmann, W. Hoffman, J. Hoffmann, M. Nüchter, B. Ondruschka and J. Filser, *Green Chem.*, 2003, **5**, 136.
15. J. Ranke, S. Stolte, R. Störmann, J. Arning and B. Jastorff, *Chem. Rev.*, 2007, **107**, 2183.
16. N. Gathergood and J. J. Scammells, *Aust. J. Chem.*, 2002, **55**, 557.
17. K. M. Docherty, R. J. Bernot, G. A. Lamberti, C. F. Kulpa and J. F. Brennecke, 1st International Congress on Ionic Liquids, Salzburg, Austria, June 19–22, 2005, Book of Abstracts, p. 264.
18. M. Freemantle, *Chem. Eng. News*, August 1, 2005, 33..
19. C. Pretti, C. Chiappe, D. Pieraccini, M. Gregori, F. Abramo, G. Monni and L. Intorre, *Green Chem.*, 2006, **8**, 238.
20. M. J. Earle, P. B. McCormac and K. R. Seddon, *Green Chem.*, 1999, **1**, 23.

21. E. B. Carter, S. L. Culver, P. A. Fox, R. D. Goode, I. Ntai, M. D. Tickell, R. K. Traylor, N. W. Hoffman and J. H. Davis, Jr., *Chem. Commun.*, 2004, 630.
22. J. M. Crostwaite, L. J. Ropel, J. L. Anthony, S. N. V. K. Aki, E. J. Maginn and J. F. Brennecke, in *Ionic Liquids IIIA: Fundamentals, Progress, Challenges, and Opportunities. Properties and Structure*, ed. R. D. Rogers and K. R. Seddon, ACS Symposium Series 901, American Chemical Society, Washington DC, 2005, p. 292.
23. P. Stepnowski, in *Ionic Liquids IV Not Just Solvents Anymore*, ed. J. F. Brennecke, R. D. Rogers and K. R. Seddon, ACS Symposium Series 975, American Chemical Society, Washington DC, 2005, p. 10.

Electrochemistry

7.1 INTRODUCTION

The appeal of ionic liquids for electrochemical applications hinges on two key features of the liquids: their ionic conductivity and their electrochemical stability.

Whereas aqueous electrolytes rely on dissolved salts for their ionic conductivity, ionic liquids exhibit intrinsic ionic conductivity because, by definition, they consist of ions. The intrinsic ionic conductivity of an ionic liquid depends on the availability of these ions to become charge carriers and on their mobility in the ionic liquid. The availability of ions is reduced by ion–ion interactions that form ion-pairs and ion aggregates that are neutral, and therefore not charge carriers.

In general, the mobility of charge carriers correlates with their rate of diffusion in the electrolyte. The diffusion rate is inversely proportion to the viscosity of the electrolyte. Ionic liquids are generally more viscous than other types of electrolyte but, on the other hand, contain more charge carriers. The viscosities of ionic liquids also tend to decrease markedly with increasing temperature. In general, the rate of diffusion of charge carriers in ionic liquids is lower than in the conventional aprotic organic solvents, such as acetonitrile, that are used in electrochemical experiments.

The nature of both the cations and anions in ionic liquids influences both their viscosities and rates of diffusion. Ionic liquids with bulky cations, for example, are more viscous than ionic liquids with smaller more mobile cations. They therefore tend to have lower conductivities. For instance, the ionic conductivity of $[C_2mim][NTf_2]$ is $9.2\,mS\,cm^{-1}$

An Introduction to Ionic Liquids
By Michael Freemantle
© Michael Freemantle 2010
Published by the Royal Society of Chemistry, www.rsc.org

compared with $2.2\,\mathrm{mS\,cm^{-1}}$ for $[C_6mim][NTf_2]$. Overall, the ionic conductivities of ionic liquids are lower than those of conventional aqueous electrolyte solutions used in electrochemistry but similar to those of solutions of inorganic electrolytes in aprotic organic solvents.

In the case of chloroaluminate room-temperature ionic liquids, addition of co-solvents such as benzene, dichloromethane or acetonitrile results in higher conductivities and lower viscosities than those of chloroaluminate melts without the cosolvents.[1] The changes are thought to arise from ion solvation leading to a decrease in the amount of ion-pairing or ion aggregation.

The electrochemical stability of an electrolyte is expressed as an electrochemical potential window. This is the window of potentials where the electrolyte is resistant to electrochemical reduction at a cathode and oxidation at an anode. Ionic liquids typically have electrochemical windows of between 2.0 and 6.0 V. For example, $[C_2mim][NTf_2]$ and $[C_4mim][PF_6]$ have windows of 4.5 and 4.2 V, respectively.

The sizes of the windows are influenced by the presence of electroactive impurities such as halide ions and the choice of working electrodes, *i.e.* the electrodes where reduction or oxidation takes place. In general, however, the stability windows of ionic liquids are better than those for aqueous electrolytes or solutions of electrolytes in organic solvents.

These two properties of ionic liquids – ionic conductivity and electrochemical stability – combined with other properties, such as tuneability of the ions, low volatility, and high thermal stability, make ionic liquids attractive as potential electrolytes for batteries, solar cells, fuel cells and other electrochemical devices and also as potential solvents for electrochemical redox reactions.

There is, however, a trade-off between electrochemical stability and conductivity. For example, the electrochemical windows of tetraalkylammonium, dialkylpyrrolidinium and dialkylpiperidiniuim ionic liquids are generally wider than those of imidazolium and sulfonium ionic liquids.[2] On the other hand, imidazolium and sulfonium ionic liquids exhibit superior conductivities.

7.2 CYCLIC VOLTAMMETRY

Various forms of voltammetry are used to study the electrochemical characteristics of ionic liquids and the electrochemical behaviour of redox active compounds in ionic liquids.

The most common method is cyclic voltammetry, whereby the potential of a working electrode is scanned to increasingly positive (anodic) potentials and then to increasingly negative (cathodic) potentials. Oxidation of the anions in an ionic liquid takes place at the anode (the positive electrode) and reduction of cations at the cathode (the negative electrode).

Scans of ionic liquids to positive and then negative potentials yield current threshold values. The flow of electrons (the current) from the anions to the anode peaks at the oxidation potential limit. This current peak is known as the oxidation peak. The current from the cathode to the cations peaks at the reductive potential limit. This is the reduction peak.

The difference between the two potential limits is the electrochemical potential window of the ionic liquid. Kinetic parameters such as diffusion coefficients of electroactive species in an ionic liquid can also be calculated from cyclic voltammetric data.

The use of acidic and basic chloroaluminate room-temperature ionic liquids as solvents for voltammetric investigations of electrochemical reactions dates back to the early 1990s. For example, in 1994 Carter and Osteryoung used the technique to study the electrochemistry of tetrathiafulvalene (**7.1**, TTF) in $[C_2mim]Cl\text{-}AlCl_3$.[3] They showed that in acidic ($AlCl_3$-rich) and basic (chloride-rich) melts of the binary mixture the compound is oxidized in two consecutive one-electron steps to TTF^+ and then to TTF^{2+}.

7.1

Impurities in an ionic liquid can influence its electrochemical window. Water in particular has a pronounced impact on the window. In 2000, cyclic voltammograms recorded by Marken and co-workers showed that water absorbed into imidazolium ionic liquids in a controlled manner resulted in a "dramatic narrowing" of the potential window.[4] For example, the potential window of dry $[C_4mim][BF_4]$ is almost twice that of wet $[C_4mim][BF_4]$ (Figure 7.1).

The researchers also investigated the impact of water on the voltammetry of redox systems dissolved in ionic liquids. One such system is the reduction of anionic hexacyanoferrate(III) ($[Fe(CN)_6]^{3-}$, also known as ferricyanide) to hexacyanoferrate(II) ($[Fe(CN)_6]^{4-}$, ferrocyanide) in the room-temperature ionic liquid 1-methyl-3-[2,6-(*S*)-dimethyloct-2-en-8-yl]imidazolium tetrafluoroborate (**7.2**). The peak current for the

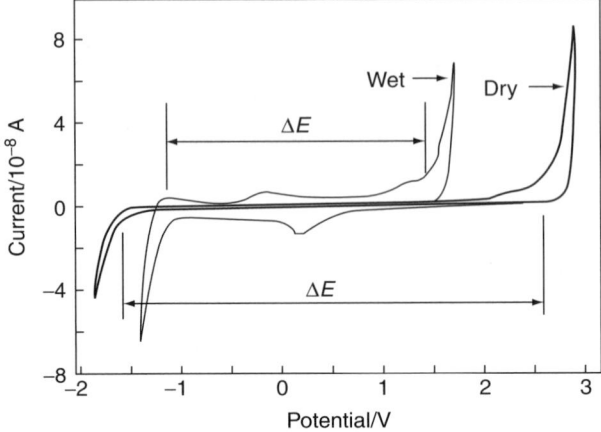

Figure 7.1 Cyclic voltammograms of dry and wet [C₄mim][BF₄] show narrowing of the electrochemical potential window (Δ*E*) when the ionic liquid is wet. (Adapted from Marken.⁴)

system increases significantly when water is present. The researchers concluded that diffusion of the ferricyanide anion is strongly inhibited by the ionic liquid, but when water is present the rate of diffusion increases dramatically.

7.2

The electrochemical potential windows of organic solvents such as acetonitrile can be extended by the use of supporting electrolytes. Tetra-butylammonium perchlorate (TBAP) is commonly used for this purpose. The compound, however, is a white powder with a melting point of 212 °C. It is therefore not a room-temperature ionic liquid.

Cyclic voltammograms recorded by Compton and co-workers reveal that room-temperature ionic liquids with trifluorotris(pentafluoroethyl) phosphate (FAP) cations are suitable replacements for TBAP as supporting electrolytes in acetonitrile solutions.⁵ The voltammograms show that FAP-based ionic liquids are better than TBAP at extending the windows of acetonitrile solutions in the oxidative (anodic) region. FAP-based ionic liquids are therefore potentially useful for voltammetric studies of compounds that oxidize at high potentials.

Compton's group has also used cyclic voltammetry to show that the electrochemistry of NaNO₃ and KNO₃ dissolved in [C₂mim][NTf₂] at ambient temperatures is comparable with the electrochemistry of the nitrate melts at temperatures in excess of 350 °C.[6] [C$_2$mim][NTf$_2$] was selected because it has an electrochemical window in excess of 5.5 V, which allows the concentrations and diffusion coefficients of both nitrates to be easily calculated. The work could prove useful for the analytical determination of nitrates, according to the authors.

In a separate study, Compton's team used cyclic voltammetry to study the electrode kinetics and mechanism of the reduction of iodine in [C$_4$mim][NTf$_2$].[7] Data extracted from voltammograms of the reduction enabled kinetic parameters such as diffusion coefficients and rate constants to be determined.

Cyclic voltammetry and other forms of voltammetry have also been used to study the electrochemical behaviour of the ferrocenium/ferrocene redox couple in various ionic liquids. Ferrocene is used as an internal standard for reporting electrode potentials in aqueous and non-aqueous solvents. In haloaluminate ionic liquids, the electrochemical behaviour of ferrocene is complicated as it depends on the acidity of the ionic liquid.

In non-haloaluminate ionic liquids, such as [C$_2$mim][BF$_4$], the ferrocenium/ferrocene redox couple is chemically stable and electrochemically reversible. In one investigation of this type of system, cyclic voltammetry and other forms of voltammetry were used to study the redox couple in three ionic liquids: [C$_4$mim][OTf], [C$_4$mim][NTf$_2$] and [C$_4$mim][PF$_6$].[8] The study showed that the redox couple is electrochemically reversible in all three ionic liquids.

7.3 BATTERIES

Interest in the use of ionic liquids with low melting points as electrolytes for batteries dates back to the early 1960s when researchers at the US Air Force Academy in Colorado Springs began investigating the possibility of using low melting point chloroaluminates such as [C$_2$mim]Cl-AlCl$_3$ and [C$_4$py]Cl-AlCl$_3$ as replacements for the high melting point molten salt electrolytes used in thermal batteries.

Thermal batteries are used for military applications such as missiles and artillery shells and also in spacecraft. They employ a molten salt, typically a eutectic mixture of LiCl and NaCl, as an electrolyte. Because the molten salt is more highly conducting than conventional battery electrolytes, they provide better energy and power density than their conventional counterparts. The electrochemical windows of molten salts such as the LiCl/NaCl eutectic are also wider than those of aqueous electrolytes. However, the eutectic has a melting point of around 355 °C.

Batteries with this electrolyte therefore require temperatures of 375–550 °C to operate. Such temperatures cause problems with the materials inside the battery and are incompatible with nearby devices.

The US Air Force Academy investigations led to the development of a molten salt battery based on what is known as the DIME (Dual Intercalating Molten Electrolyte) concept.[9] Anions of an ionic liquid electrolyte are intercalated into a carbon electrode and the cations into another carbon electrode. Ionic liquids such as $[C_2mim][BF_4]$ and $[C_2mim][PF_6]$ were shown to be effective as electrolytes for these types of batteries.

Ionic liquids also show potential for enhancing the performance of the lithium-based rechargeable batteries that are widely used in portable electronic devices such as laptop computers and cell phones. Lithium-ion batteries exhibit the highest energy density among rechargeable energy storage systems.

The batteries typically contain a graphite negative electrode (the anode) and a positive electrode (the cathode) consisting of a lithium transition metal oxide such as $LiCoO_2$ or $LiNiO_2$. The batteries employ an electrolyte mixture of an electrochemically stable lithium salt, *e.g.* $LiPF_6$, and ethylene carbonate dissolved in an organic solvent such as dimethyl carbonate. The ethylene carbonate forms a passivation film on the anode that protects the electrolyte against reduction while it is ionically conductive. The film ensures a high number of charge and discharge cycles during which lithium ions respectively intercalate into and de-intercalate from the graphite electrode. The ions travel from the cathode to the graphite anode when charging and from the anode to the cathode during discharge. A drawback of these batteries is the low boiling point and easy flammability of the organic solvent, which poses a safety risk. The volatility of the solvent reduces the temperature range for applications of these batteries.

Ionic liquids, in contrast, are fluid over a wider range of temperatures and much less flammable than organic solvents. As they also exhibit desirable electrochemical properties, ionic liquids have solicited considerable interest as replacements for organic solvents in lithium-ion batteries. Consequently, various lithium-ion batteries that use ionic liquid/lithium salt mixtures as electrolytes have been devised in recent years.

One such battery uses an electrolyte consisting of a lithium salt dissolved in the ionic liquid $[C_2mim][NTf_2]$. Lithium ions have been shown to reversibly intercalate into a standard graphite electrode when vinylene carbonate is added in small amounts to the ionic liquid/lithium salt electrolyte.[10] The cycling behaviour of the graphite electrode in the ionic liquid based electrolyte is comparable to that of a graphite electrode in a standard electrolyte based on organic carbonates. This means that the capacity of the battery with the ionic liquid based electrolyte to charge and discharge reversibly is comparable to that of the battery with the standard electrolyte.

A lithium-ion battery with a lithium metal anode, a cathode consisting of $LiCoO_2$ coated with ZrO_2, and an ionic liquid/lithium salt electrolyte has also been fabricated and tested.[11] The room-temperature ionic liquid used for the battery consists of *N,N*-diethyl-*N*-methyl-*N*-(2-methoxyethyl)ammonium cations and $[NTf_2]^-$ anions. The cathode is coated with ZrO_2, a stable oxide, to improve the charge–discharge performance of the battery over repeated cycles.

One of the limitations of using ionic liquids as electrolytes in lithium-ion batteries, however, is the competition between the lithium ions and the ionic liquid cations for the anions of the ionic liquids.[12] The small lithium ions tend to form anionic complexes and therefore lose their mobility.

Polymer composite gel electrolytes consisting of a polymer and an ionic liquid have also shown promise for use in lithium-ion batteries. Sutto and co-workers demonstrated that a gel electrolyte composed of polyvinylidenedifluoro-hexafluoropropylene and an ionic liquid with 1,2-dimethyl-3-alkylimidazolium cations and $[NTf_2]^-$ anions performs as well as the more common organic electrolytes when lithium metal is used as an anode and a metal oxide such as $LiMn_2O_4$ is used as a cathode.[13]

7.4 FUEL CELLS

Protic ionic liquids are potentially useful as anhydrous proton-conducting electrolytes for hydrogen fuel cells. The ionic liquids are also known as Brønsted acid–base ionic liquids as they are formed by the protonation of a Brønsted base, typically an organic amine, by a Brønsted acid.

A fuel cell is a type of primary cell in which the reactants are continuously replaced as they are consumed and the products are continuously removed. A simple hydrogen/oxygen fuel cell has two porous carbon electrodes containing a platinum catalyst. The electrodes are separated by a proton-conducting electrolyte. Hydrogen (the fuel) and oxygen (the oxidant) are fed to the anode and cathode, respectively. The hydrogen diffuses through the anode, which catalyses its oxidation to protons and electrons. The electrons move into the external circuit. The electrolyte conducts the protons to the cathode catalyst where they reduce the oxygen to produce water.

In 2003, Watanabe and co-workers described a novel family of proton-conducting ionic liquids that might serve as protic electrolytes for hydrogen/oxygen fuel cells operating at temperatures above $100\,^{\circ}\text{C}$.[14] The Brønsted acid–base ionic liquids are prepared by combining a super strong acid, bis(trifluoromethanesulfonyl)amide, with various nitrogen-containing organic compounds under solvent-free conditions. The group have shown that an ionic liquid derived from equimolar amounts of the amide and imidazole could be used as an electrolyte in a fuel cell. When two of the fuel cells are connected in series, they generate sufficient power to operate a calculator.

Angell and co-workers have investigated the function of binary mixtures of inorganic salts with protonated cations as protic electrolytes for fuel cells operating in the temperature range 100–$200\,^{\circ}\text{C}$.[15] The melts studied consist of two ammonium salts with eutectic temperatures below $110\,^{\circ}\text{C}$. They were found to exhibit higher conductivities and better stabilities as fuel cell electrolytes than most aprotic ionic liquids. The most suitable for fuel cell purposes was a binary mixture of ammonium trifluoroacetate and ammonium nitrate, which has a eutectic temperature of $79\,^{\circ}\text{C}$. The mixture was the most stable of all the electrolytes tested and yielded a cell voltage of $1.2\,\text{V}$.

7.5 DYE-SENSITIZED SOLAR CELLS

The possibility of using ionic liquids as electrolytes in dye-sensitized solar cells (DSCs) has been an active area of research since the early 2000s. DSCs are photovoltaic cells that mimic photosynthesis by using a photosensitive dye, typically a ruthenium bipyridyl complex. The dye has the same role as chlorophyll in plants. It harvests light. The potential low cost of fabricating DSCs and the flexibility of the devices make them promising alternatives to conventional solar cells based on crystalline silicon.

DSCs typically consist of a fluorine-doped SnO_2-conducting glass counter electrode and semiconductor titanium dioxide nanoparticles dispersed in a film of a liquid or solid electrolyte. The photosensitive dye is coated on the TiO_2 particles. Incident photons excite electrons in the dye. The electrons are then injected into the conduction band of the TiO_2. The positive holes left in the dye are carried away by an electrolyte to the counter electrode.

A typical DSC electrolyte consists of iodide–triiodide (I^-/I_3^-) redox couple dissolved in an organic solvent such as acetonitrile. However, as the organic solvent is volatile, solar heating of the cells can result in practical problems for the construction and durability of the devices. For example, the cells require good sealing to prevent the solvent leaking and damaging the plastic components. For that reason, ionic liquids, because they are non-volatile, are seen as possible solvent-free electrolytes for these cells.

Many of the investigations have been carried out by a team led by Grätzel, who was one of the inventors of DSCs.[16] The studies have focused principally on the use of ionic liquids with iodide anions. Such ionic liquids can be used as a solvent for the dye and their anions as a source of iodide for the redox couple.

In an early study, reported in 2002, Grätzel's group demonstrated that a quasi-solid-state dye-sensitized nanocrystalline TiO_2 solar cell could be fabricated using an ionic polymer gel, containing $[C_3mim]I$ and a polymer, as the electrolyte.[17] The cell yielded an absorbed light-to-energy conversion efficiency of just over 5%, which is similar to the level of efficiency of the photovoltaic cells that power calculators.

The group subsequently used electrochemically stable silica nanoparticles to gel the same ionic liquid. The quasi-solid-state DSC containing this gel electrolyte achieved 7% efficiency.[18]

In 2003, Grätzel and co-workers reported a comparable solar-to-electricity efficiency for a DSC with an ionic liquid electrolyte consisting of $[C_3mim]I$, $[C_2mim][dca]$ and lithium iodide.[19]

The relatively high viscosities of pure ionic liquids containing iodide and triiodide ions impose performance limitations when they are used as solvent-free redox electrolytes in solar cells. Ionic liquids with low viscosities and other types of anions have therefore been tested as solvent-free electrolytes for DSCs.

Grätzel's group, for example, has explored the possibility of using an ionic liquid electrolyte based on the pseudohalide redox couple $SeCN^-/(SeCN)_3^-$. In 2004, the group showed that "unprecedented" solar-to-electric power conversion efficiencies for solvent-free redox electrolytes could be achieved using the low viscosity ionic liquid 1-ethyl-3-methylimidazolium selenocyanate, $[C_2mim][SeCN]$, as a solvent-free ionic liquid electrolyte.[20] At room temperature, pure $[C_2mim][SeCN]$ has a viscosity of 25 cP, compared with 880 cP for $[C_3mim]I$. DSCs with the $[C_2mim][SeCN]$ electrolyte achieved power conversion efficiencies of 7.5–8.3%.

$[C_2mim][SeCN]$ and other low-viscosity ionic liquid redox electrolytes, however, suffer from a significant drawback. They tend to be unstable when subjected to elevated temperatures and prolonged exposure to sunlight.[21]

By 2008, the best reported performance, in terms of conversion efficiency and stability, for a DSC based on a solvent-free electrolyte had been achieved using a eutectic melt of three imidazolium ionic liquids: [C_1mim]I, [C_2mim]I and 1-allyl-3-methylimidazolium iodide.[21] A mixture of the three ionic liquids in a molar ratio of 1 : 1 : 1 has a melting point below 0 °C and a "strikingly" high room-temperature conductivity according to the authors of the report. A DSC based on the electrolyte exhibits 8.2% efficiency and "excellent" stability in full sunlight, the authors observed.

Whereas the best conversion efficiencies for DSCs with solvent-free electrolytes are around 8%, conversion efficiencies of up to 11% have been achieved for DSCs with electrolytes dissolved in organic solvents such as acetonitrile.

In 2007, Grätzel and co-workers showed that a combination of an ionic liquid and an organic solvent could be used to prepare a non-volatile I^-/ I_3^--based electrolyte for a high-efficiency DSC.[22] The group achieved photoelectrical conversion efficiencies of up to 9.5% using a new type of ruthenium bipyridyl dye and an electrolyte that contained the ionic liquid [C_3mim]I and the organic solvent 3-methoxypropionitrile. DSCs with this non-volatile organic solvent–ionic liquid electrolyte also exhibited unprecedented long-term stability of around 1000 hours.

7.6 ELECTRODEPOSITION

Electrodeposition is an electrolytic process that deposits an electropositive material, typically a metal or an alloy, on a cathode. The electrolyte contains cations of the material to be deposited. These are reduced at the cathode.

Electrodeposition is used for electroplating, a process that deposits a thin layer of material with desirable properties, such as corrosion resistance, onto a cathode, which is the object to be coated. Electrodeposition is also used to extract metals from ores in a process known as electrowinning or electroextraction. The electrolyte is either a molten form of the ore or a solution of the ore. Finally, electrodeposition is used for electrorefining impure metals. The anode consists of the impure metal. The pure metal deposits on the cathode.

It is not possible to use aqueous solutions for the electrodeposition of highly electropositive metals such aluminium, lithium, and magnesium because hydrogen is evolved at the cathode. Electrodeposition of reactive metals therefore requires an aprotic electrolyte.

The wide electrochemical windows, thermal stability, good ionic conductivity, anhydrous nature and negligible vapour pressures, combined with the ability to dissolve metal salts, make ionic liquids good candidates for the electrodeposition of not only highly electropositive elements but also less reactive elements that can be deposited from aqueous solutions.

The potential use of ionic liquids as electrolytes for the electrodeposition of metals was recognized over 50 years ago when Hurley and Wier showed that various metals could be deposited on a platinum cathode from fused mixtures of ethylpyridinium bromide and metal chlorides using a carbon rod anode.[23]

The two chemists were particularly interested in finding less expensive and more facile ways of extracting aluminium by electrolysis. The traditional industrial aluminium extraction process is highly energy intensive, employs a solution of alumina, Al_2O_3, in cryolite, Na_3AlF_6, a salt that melts at 970 °C. Hurley and Wier found that a mixture of aluminium chloride and 1-ethylpyridinium bromide in the molar ratio 2:1 has a eutectic temperature of −40 °C. They showed that the melt could be used as an electrolyte to deposit aluminium on a cathode made of iron, steel or some other metal or alloy. The anode was made of aluminium.

Since then, numerous studies have been carried out on the use of chloroaluminate ionic liquids for the low-temperature electrodeposition of metals and their alloys, particularly those elements that cannot be deposited from aqueous solutions and could previously only be deposited from high-temperature molten salts. Various research groups have reported the use of chloroaluminate ionic liquids for the electroposition of elements such aluminium, sodium, lithium and iron, and alloys, particularly aluminium alloys.[24] These metals and alloys cannot be electrodeposited from water because of their reactivity. Chloroaluminate ionic liquids have also been used to deposit metals that can also be obtained by electrodeposition from aqueous electrolytes. These include copper, silver, nickel, gold, palladium, zinc and tin.

Research has also been carried out on the use of non-chloroaluminate ionic liquids as electrolytes for electrodeposition. For example, the electrodeposition of the semiconducting element germanium on gold films from $[C_4mim][PF_6]$ saturated with $GeCl_4$ has been reported.[25]

$[C_4mim][PF_6]$, however, is air stable but not water stable. In recent years, there have been several reports on the use of air- and water-stable ionic liquids as electrolytes for electrodeposition of elements. Aluminium, for example, can be electrodeposited on steel using a mixture of the air- and water-stable ionic liquid $[C_2mim][NTf_2]$ and $AlCl_3$.[26]

An air- and water-stable ionic liquid containing 1-butyl-1-methyl-pyrrolidinium cations and [NTf$_2$]$^-$ anions has also proved useful for depositing various elements, notably silicon. This semiconducting element has a very negative reduction potential and therefore cannot be deposited from aqueous solutions because of hydrogen evolution at the cathode. Molinari and co-workers have shown that a solution of the ionic liquid and SiCl$_4$ can be used as an electrolyte to grow silicon nanowires with controlled diameters.[27] An insulating polycarbonate nanoporous membrane is used as a template to control the diameters of the nanowires.

Eutectic mixtures of choline chloride, [HOCH$_2$CH$_2$N(CH$_3$)$_3$]Cl, and metal salts have been employed for the electrodeposition of metals. For example, a zinc chloride/choline chloride eutectic has been shown to have analogous electrochemical properties to chloroaluminate melts.[28] Unlike chloroaluminate melts, the choline chloride eutectic is stable in air and water. It is also easy to prepare. The eutectic can be used to deposit a corrosion resistant non-porous coating of zinc on a steel electrode.

Another investigation showed that chromium can be efficiently electrodeposited onto stainless steel from a eutectic of choline chloride and hydrated chromium(III) chloride, CrCl$_3$.6H$_2$O, mixed in the molar ratio of 1 : 2.[29] Addition of LiCl to the eutectic electrolyte results in the deposition of black nanocrystalline chromium films that are corrosion resistant and crack free. Conventional chromium plating uses an aqueous electrolyte containing a Cr(VI) salt that is highly toxic and carcinogenic. Hydrogen is also evolved. Cr(III) salts, on the other hand, are significantly less toxic than Cr(VI) salts. In addition, the use of the eutectic and Cr(III) salt is more efficient, resulting in reduced power consumption. As water is not used in the process, hydrogen evolution is negligible.

Titanium, because of its mechanical strength and resistance to corrosion, is potentially one of the most attractive elements for electrodeposition at low temperatures. However, attempts to electrodeposit the element from its halides, such as TiCl$_4$, in various ionic liquids have so far proved unsuccessful as it appears that it is not possible to reduce Ti^{4+} ions completely to the metal in the presence of the halide.[30]

Ionic liquids have been employed not only as electrolytes for electrodeposition processes such as electroplating but also for electropolishing, which is the opposite of electroplating. With electropolishing, a metal is removed, whereas electroplating deposits a metal. The metal to be polished is oxidized at the anode.

Abbott and co-workers have shown that a eutectic based on an ionic liquid composed of ethylene glycol, $HOCH_2CH_2OH$ and choline chloride, offers a practical alternative to the use of mixtures of phosphoric acid and sulfuric acid that are often used for electropolishing steel.[31] The choline chloride eutectic mixture reacts with the metal cations on the surface of the steel anode and forms an insoluble complex precipitate that sinks to the bottom of the electropolishing bath. The cell liquid is filtered periodically and the metal complex is collected. The eutectic fluid and metal can then be recycled.

7.7 ACTUATORS

The possibility of using ionic liquids in combination with electroactive polymers for the development of light weight electrochemical mechanical actuators has excited the attention of several research groups over recent years. These actuators typically rely on the application of a voltage to a π-conjugated conducting polymer, such as polyaniline (**7.3**), polypyrrole (**7.4**) or polythiophene (**7.5**), to activate a change in its dimensions. When the voltage is switched off, the polymer reverts to its original shape. These physical changes simulate those that occur in human muscles and could potentially be exploited for robotic applications and the fabrication of artificial muscles that can replace damaged human muscles.

7.3   **7.4**   **7.5**

Electroactive polymer actuators depend on electrochemical oxidation and reduction for their mechanical action. They therefore require an electrolyte. As the voltage is switched on and off over a number of cycles, ions and solvent molecules move reversibly between the electrolyte and the polymer. Most studies have been carried out using aqueous electrolytes, which, unlike ionic liquids, are not only volatile but also have narrow electrochemical windows.

In 2002, Mattes and co-workers described the fabrication of polyaniline fibre and yarn actuators and polypyrrole tube actuators with $[C_4mim][BF_4]$ and $[C_4mim][PF_6]$ electrolytes.[32] They showed that the polymers could be electrochemically cycled with fast cycle switching speeds for up to one million cycles without actuator failure.

   In a related development, Ricks-Laskoski and Snow demonstrated the actuation of an ionic liquid polymer by means of a process known as electrowetting.[33] The process relies on the ability of a liquid droplet to spread on a surface of a substrate when a voltage is applied to the droplet.

   The two chemists prepared the ionic polymer **7.6** from the crystalline compound 2-acrylamido-2-methyl-1-propanesulfonic acid. Addition of tris[2-(2-methoxyethoxy)ethyl]amine converts the compound into a liquid ammonium salt that is then polymerized to the ionic liquid polymer by the addition of a free radical initiator. The polymer has a glass-transition temperature of −47 °C. When a voltage is applied to a droplet of the polymer on a substrate the surface tension and contact angle of the polymer droplet at the solid–liquid interface decrease, resulting in spreading of the liquid.

7.6

7.8 SUPERCAPACITORS

A supercapacitor is a device that stores electrical energy by accumulating electric charges at the electric double layer at the interface between an electrode and an electrolyte when a voltage is applied across a pair of electrodes. The devices, also known as electrochemical double layer capacitors or electric double layer capacitors, typically employ an electrolyte solution of a salt dissolved in a polar aprotic organic solvent and activated carbon electrodes with high surface areas.

   Unlike batteries, which store energy chemically by redox reactions, supercapacitors store energy physically by the movement of ions through the electrolyte. They can therefore be rapidly charged or discharged. In general, supercapacitors have better power densities (power-to-weight ratios) than batteries and conventional capacitors because of various factors, including the high surface areas of their electrodes, fast kinetics and lack of redox reactions.

In 1990, Tanahashi and co-workers investigated the characteristics of supercapacitors that used an electrolyte consisting of the ionic compound tetraethylammonium tetrafluoroborate, $[N_{2\,2\,2\,2}][BF_4]$, dissolved in propylene carbonate, a polar aprotic solvent.[34] Since then, commercial supercapacitors have been developed that use organic electrolyte solutions containing $[N_{2\,2\,2\,2}][BF_4]$ or tetraethylphosphonium tetrafluoroborate, $[P_{2\,2\,2\,2}][BF_4]$.[35] However, salts such as $[N_{2\,2\,2\,2}][BF_4]$ are solid at room temperature and have relatively high melting points. They cannot therefore be classified as ionic liquids. They also decompose when they melt.

Ionic liquids have been under investigation as potential electrolytes for supercapacitors for more than a decade. In 1997, McEwen and co-workers reported that $[C_2mim][PF_6]$ and $[C_2mim][BF_4]$ were suitable electrolytes for the devices.[36] However, once again, the electrolytes were dissolved in organic solvents such as propylene carbonate.

Two years later, McEwen and colleagues described an investigation into the electrochemical properties of various imidazolium ionic liquids and showed that many of the liquid salts could be used as electrolytes for supercapacitors not only when dissolved in organic solvents but also when neat, *i.e.* in the absence of a solvent.[37]

In 2004, Sato and co-workers described the preparation of two novel quaternary ammonium room-temperature ionic liquids for supercapacitor applications.[38] The salts, which contain a methoxyethyl group on the nitrogen atom of the cation, are [DEME][BF_4] and [DEME][NTf_2], where DEME stands for *N,N*-diethyl-*N*-methyl-*N*-(2-methoxyethyl)ammonium. They selected [DEME][BF_4] to fabricate a supercapacitor because it has not only high conductivity but also a wider potential window than [DEME][NTf_2]. At temperatures over 100 °C, the supercapacitor has a higher capacity and better charge–discharge durability than conventional supercapacitors with an organic liquid electrolyte such as $[N_{2\,2\,2\,2}][BF_4]$ in propylene carbonate. However, below 40 °C, its capacity was inferior to that of conventional supercapacitors because of the high viscosity of the ionic liquid.

Mastragostino and co-workers subsequently showed that an ionic liquid with *N*-butyl-*N*-methylpyrrolidinium cations and [NTf_2]⁻ anions could be used as a "solvent-free green electrolyte" for a "hybrid" supercapacitor.[39] Their supercapacitor has hybrid electrodes consisting of a composite of activated carbon and poly(3-methylthiophene). The device could potentially be used to store the energy produced by fuel cells, according to the researchers. The group reported that it performed well over several thousand charge–discharge cycles at temperatures required for coupling directly to fuel cells.

In another development, a team led by Watanabe showed that an
ionic liquid can be used not only as an electrolyte for a supercapacitor
but also as part of its electrodes.[40] The electrodes were prepared by
grinding up single-walled carbon nanotubes with [C$_2$mim][NTf$_2$] to form
what they called a "bucky gel". [C$_2$mim][NTf$_2$] was also used as the
electrolyte. The team found that the bucky gel electrodes enhanced the
performance of the supercapacitor compared with one that used
conventional activated carbon electrodes and [C$_2$mim][NTf$_2$] as the
electrolyte.

7.9   ELECTROSYNTHESIS

The synthesis of compounds by application of a voltage to a solution of
starting materials and an electrolyte in an electrochemical cell is known
as electrosynthesis.

Much of the early work on the use of ionic liquids as electrolytes for
electrosynthesis focused on the synthesis of conducting polymers. In the
early 1980s, Pickup and Osteryoung showed that haloaluminate ionic
liquids could be used as electrolytes for the synthesis of films of redox
polymers on anodes by electrochemical oxidative dehydrogenation of
the monomers. For example, they synthesized polypyrrole films by the
anodic oxidative polymerization of pyrrole in neutral melts of [C$_4$py]Cl-
AlCl$_3$ at ambient temperatures.[41] Cyclic voltammetry revealed that the
films were electrochemically similar to polypyrrole films prepared in
acetonitrile solutions. The films are conductors when oxidized and are
potentially useful electrode materials.

Osteryoung was interested in the use of conducting polymers such as
polypyrrole and polythiophene as charge-storing materials for positive
electrodes (cathodes) in rechargeable batteries with room-temperature
ionic liquid electrolytes. He used not only pyridinium but also imidazolium haloaluminate ionic liquids for electrosynthesis of the polymers.
In 1987, he reported work with Janiszewska on the electrochemistry of
polythiophene films in [C$_2$mim]Cl-AlCl$_3$ mixtures.[42] Cyclic voltammetry
showed that the films are conductive in the oxidized state and non-
conductive when reduced. The films are also more stable than those
prepared in acetonitrile solutions.

Non-haloaluminate ionic liquids have also been used as electrolytes for
the electrosynthesis of conducting polymers. Mattes and co-workers have
synthesized poly(3,4-ethylenedioxythiophene) (7.7), commonly known as
PEDOT, and polyaniline from their monomers in [C$_4$mim][BF$_4$] electrolytes.[43] They showed that the two polymers could be used to fabricate

electrochromic numeric displays. Electrochromic devices use materials that change colour or opacity when a voltage is applied to them.

7.7

The use of ionic liquids as electrolytes for electrosynthesis is not confined to the synthesis of conducting polymers. Lu and co-workers have developed an "efficient and environmentally benign" electrochemical procedure for synthesizing dialkyl carbonates, such as dimethyl carbonate, using $[C_4mim][BF_4]$ as an electrolyte.[44] The esters are used in the chemical and pharmaceutical industries but procedures for their syntheses are not environmentally friendly. One route requires the use of phosgene ($COCl_2$), a toxic and corrosive compound.

Lu and colleagues showed that good yields of various dialkyl carbonates could be obtained under mild conditions by the electrochemical reduction of CO_2 at a cathode using primary and secondary alcohols and an alkyl iodide as an alkylating agent. The ionic liquid was saturated with CO_2.

REFERENCES

1. R. L. Perry, K. M. Jones, W. D. Scott, Q. Liao and C. L. Hussey, *J. Chem. Eng. Data*, 1995, **40**, 615.
2. T. Tsuda and C. L. Hussey, *Electrochem. Soc.: Interface*, Spring 2007, 42.
3. M. T. Carter and R. A. Osteryoung, *J. Electrochem. Soc.*, 1994, **141**, 1713.
4. U. Schröder, J. D. Wadhawan, R. G. Compton, F. Marken, P. A. Z. Suarez, C. S. Consorti, R. F. de Souza and J. Dupont, *New J. Chem.*, 2000, **24**, 1009.
5. M. C. Buzzeo, C. Hardacre and R. G. Compton, *ChemPhysChem.*, 2006, **7**, 176.
6. T. L. Broder, D. S. Silvester, L. Aldous, C. Hardacre, A. Crossley and R. G. Compton, *New J. Chem.*, 2007, **31**, 966.
7. E. I. Rogers, I. Streeter, L. Aldous, C. Hardacre and R. G. Compton, *J Phys. Chem. C*, 2008, **112**, 10976.

8. M. C. Lagunas, W. R. Pitner, J.-A. van den Berg and K. R. Seddon, in *Ionic Liquids as Green Solvents: Progress and Prospects*, ed. R. D. Rogers and K. R. Seddon, ACS Symposium Series 856, American Chemical Society, Washington DC, 2003, p. 421.

9. J. S. Wilkes, in *Green Industrial Applications of Ionic Liquids*, ed. R. D. Rogers, K. R. Seddon and S. Volkov, NATO Science Series, Kluwer Academic Publishers, Dordrecht, 2002, p. 295.

10. M. Holzapfel, C. Jost, A. Prodi-Schwab, F. Krumeich, A. Würsig, H. Buqa and P. Novák, *Carbon*, 2005, **43**, 1488.

11. S. Seki, Y. Kobayashi, H. Miyashiro, Y. Ohno, A. Usami, Y. Mita, M. Watanabe and N. Terada, *Chem. Commun.*, 2006, **1**, 544.

12. C. A. Angell, N. Byrne and J.-P. Belieres, *Acc. Chem. Res.*, 2007, **40**, 1228.

13. H. C. De Long, P. C. Trulove and T. E. Sutto, in *Ionic Liquids as Green Solvents: Progress and Prospects*, ed. R. D. Rogers and K. R. Seddon, ACS Symposium Series 856, American Chemical Society, Washington DC, 2003, p. 478.

14. M. A. B. H. Susan, A. Noda, S. Mitsushima and M. Watanabe, *Chem. Commun.*, 2003, 938.

15. J.-P. Belieres, D. Gervasio and C. A. Angell, *Chem. Commun.*, 2006, 4799.

16. B. O'Regan and M. Grätzel, *Nature*, 1991, **353**, 737.

17. P. Wang, S. M. Zakeeruddin, I. Exnar and M. Grätzel, *Chem. Commun.*, 2002, 2972.

18. P. Wang, S. M. Zakeeruddin, P. Comte, I. Exnar and M. Grätzel, *J. Am. Chem. Soc.*, 2003, **125**, 1166.

19. P. Wang, S. M. Zakeeruddin, J.-E. Moser and M. Grätzel, *J. Phys. Chem. B*, 2003, **107**, 13280.

20. P. Wang, S. M. Zakeeruddin, J.-E. Moser, R. Humphry-Baker and M. Grätzel, *J. Am. Chem. Soc.*, 2004, **126**, 7165.

21. Y. Bai, Y. Cao, J. Zhang, M. Wang, R. Li, P. Wang, S. M. Zakeeruddin and M. Grätzel, *Nat. Mater.*, 2008, **7**, 628.

22. D. Kuang, C. Klein, S. Ito, J.-E. Moser, R. Humphry-Baker, N. Evans, F. Duriaux, C. Grätzel, S. M. Zakeeruddin and M. Grätzel, *Adv. Mater.*, 2007, **19**, 1133.

23. F. H. Hurley and T. P. Wier, *J. Electrochem. Soc.*, 1951, **98**, 203–207.

24. F. Endres, *ChemPhysChem*, 2002, **3**, 144.

25. F. Endres and S. Zein El Abedin, *Chem. Commun.*, 2002, 892.

26. S. Zein El Abedin and F. Endres, *Acc. Chem. Res.*, 2007, **40**, 1106.
27. J. Mallet, M. Molinari, F. Martineau, F. Delavoie, P. Fricoteaux and M. Troyon, *Nano Lett.*, 2008, **8**, 3468.
28. A. P. Abbott, G. Capper, D. L. Davies, H. Munro, R. K. Rasheed and V. Tambyrajah, in *Ionic Liquids as Green Solvents: Progress and Prospects*, ed. R. D. Rogers and K. R. Seddon, ACS Symposium Series 856, American Chemical Society, Washington DC, 2003, p. 439.
29. A. P. Abbott, G. Capper, D. L. Davies, R. K. Rasheed, J. Archer and C. John, *Trans. Inst. Met. Fin.*, 2004, **82**, 14.
30. F. Endres, S. Zein El Abedin, A. Y. Saad, E. M. Moustafa, N. Borissenko, W. E. Price, G. G. Wallace, D. R. MacFarlane, P. J. Newman and A. Bund, *Phys. Chem. Chem. Phys.*, 2008, **10**, 2189.
31. A. P. Abbott, K. J. McKenzie and K. S. Ryder, in *Ionic Liquids IV: Not Just Solvents Anymore*, ed. J. F. Brennecke, R. D. Rogers and K. R. Seddon, ACS Symposium Series 975, American Chemical Society, Washington DC, 2007, p. 186.
32. W. Lu, A. G. Fadeev, B. Qi, E. Smela, B. R. Mattes, J. Ding, G. M. Spinks, J. Mazurkiewicz, D. Zhou, G. G. Wallace, D. R. MacFarlane, S. A. Forsyth and M. Forsyth, *Science*, 2002, **297**, 983.
33. H. L. Ricks-Laskoski and A. W. Snow, *J. Am. Chem. Soc.*, 2006, **128**, 12402.
34. I. Tanahashi, A. Yoshida and A. Nishino, *J. Electrochem. Soc.*, 1990, **137**, 3052.
35. K. Xu, M. S. Ding and T. R. Jow, *J. Electrochem. Soc.*, 2001, **148**, A267.
36. A. B. McEwen, S. F. McDevitt and V. R. Koch, *J. Electrochem. Soc.*, 1997, **144**, L84.
37. A. B. McEwen, H. L. Ngo, K. LeCompte and J. L. Goldman, *J. Electrochem. Soc.*, 1999, **146**, 1687.
38. T. Sato, G. Masuda and K. Takagi, *Electrochim. Acta*, 2004, **49**, 3603.
39. A. Balducci, W. A. Henderson, M. Mastragostino, S. Passerini, P. Simon and F. Soavi, *Electrochim. Acta*, 2005, **50**, 2233.
40. T. Katakabe, T. Kaneko, M. Watanabe, T. Fukushima and T. Aida, *J. Electrochem. Soc.*, 2005, **152**, A1913.
41. P. G. Pickup and R. A. Osteryoung, *J. Am. Chem. Soc.*, 1984, **106**, 2294.

42. L. Janiszewska and R. A. Osteryoung, *J. Electrochem. Soc.*, 1987, **134**, 2787.
43. W. Lu, A. G. Fadeev, B. Qi and B. R. Mattes, *Synth. Met.*, 2003, **135**, 139.
44. L. Zhang, D. Niu, K. Zhang, G. Zhang, Y. Luo and J. Lu, *Green Chem.*, 2008, **10**, 202.

CHAPTER 8
Catalysis

8.1 INTRODUCTION

Over the past ten years, ionic liquids have proved increasingly versatile and attractive as media for organic synthesis in general (Chapters 10 and 11) and for catalysis in particular. They have also proved useful for biosynthesis and enzyme catalysis (Chapter 12).

The physical and chemical properties of ionic liquids, such as lack of volatility and relatively high thermal stability, often make ionic liquids more attractive as media for catalysis than water and conventional organic solvents.

The opportunity to select or design ionic liquids with specific solubility or miscibility properties is one of the most appealing features of ionic liquids. Many ionic liquids offer a highly polar but non-coordinating environment for catalysts. For example, metal complexes used in transition metal catalysis are generally poorly soluble in non-polar, non-coordinating molecular solvents such as hexane. Polar solvents such as acetonitrile, on the other hand, tend to coordinate metal complexes. In many cases, when a catalyst is dissolved in a polar solvent in sufficient concentration, the solvent coordinates on the active site and therefore blocks its activity. Ionic liquids with non-coordinating anions such as $[BF_4]^-$ can therefore, in many instances, replace polar coordinating solvents.

An Introduction to Ionic Liquids
By Michael Freemantle
© Michael Freemantle 2010
Published by the Royal Society of Chemistry, www.rsc.org

8.2   MONOPHASIC AND BIPHASIC CATALYSIS

The ability of ionic liquids to dissolve a broad range of inorganic, organic and organometallic compounds and gases such as hydrogen and oxygen, combined with their immiscibility with some organic solvents and in other cases aqueous solvents, makes them particularly attractive for both monophasic and multiphasic catalysis.

In monophasic systems, the substrate and the catalyst are dissolved in the ionic liquid. In some cases the ionic liquid is both the solvent and the catalyst or co-catalyst. With such systems, the products are separated by solvent extraction, distillation or some other means.

One of the first papers to report the use of "low-melting salts" (or room-temperature ionic liquids as they became known) as media for monophasic catalysis was by Parshall in the early 1970s. In a paper entitled "Catalysis in Molten Salt Media" he observed that separation of products from catalyst and solvent without decomposition of the catalyst is a major problem in homogeneous catalysis.[1]

Parshall reported that tetraethylammonium salts with trichlorogermanate, $[GeCl_3]^-$, and trichlorostannate, $[SnCl_3]^-$, anions melt at 68 and 78 °C, respectively, and are good solvents for olefins. He showed that $PtCl_2$ dissolved in these low-melting salts gives deep red solutions that catalyse the hydrogenation, isomerization, hydroformylation and carboalkoxylation of olefins. He separated the products of the reactions from the salt solutions by vacuum distillation. Parshall noted that the salts are not only stable but also act as ligands that stabilize the platinum complexes.

In biphasic systems, the catalyst typically dissolves in the ionic liquid and the reactants and products dissolve in a second phase, generally an organic solvent or an aqueous solution that is immiscible with the ionic liquid. The reactants are exposed to the catalyst by rapidly stirring the two phases or by sonification or emulsification of the biphasic mixture. The products in the non-ionic liquid phase can be easily separated by decantation when the reaction is complete. The ionic liquid–catalyst solution can then often be reused with little or no loss of catalytic activity.

These two-phase or "biphasic" catalytic processes offer the benefits of both homogeneous and heterogeneous catalysis. In homogeneous catalysis, the reaction occurs in the same phase as the catalyst. The reactions are generally carried out under mild conditions with high efficiency and selectivity for the desired products. In heterogeneous catalysis, the catalyst, usually a solid, is in a different phase from the reactants and products which are generally liquids or gases. The catalyst can therefore be readily separated from reaction products and used again. This allows optimal use of the catalyst.

Catalysis in ionic liquid systems, whether monophasic or biphasic, is a form of homogeneous catalysis since the catalyst is dissolved in the ionic liquid and the reaction takes place in the ionic liquid. Ionic liquids can also be exploited for heterogeneous catalysis. For example, they can be tethered to a solid support in what is known as supported ionic liquid phase catalysis (Section 8.5).

8.3 SOLVENTS, CATALYSTS AND LIGANDS

Ionic liquids can act as reaction media for homogeneous catalysis in several ways. In the simplest case, they perform as inert solvents without any obvious chemical interaction with the reactants or catalysts. In this case, they have sometimes been called "innocent" solvents. Ionic liquids may also act as solvents for the reactants in catalysis and at the same time be involved in catalytic activity as catalysts, co-catalysts, ligands or ligand precursors. In other cases, one ionic liquid is used as a catalyst for a reaction and another ionic liquid employed as the solvent.

Wasserscheid and Schulz observed that innocent ionic liquid solvents consist of inert ions, notably weakly coordinating anions such as $[BF_4]^-$, $[NTf_2]^-$ or $[PF_6]^-$, and cations that do not coordinate with the catalyst.[2] $[C_4mim][PF_6]$ and related ionic liquids with longer alkyl chains on the imidazolium cation are examples. These ionic liquids are weakly co-ordinating but good solvents for cationic transition metal complexes.

In 2001, Wasserscheid, Gordon and others described the use of innocent ionic liquids such as $[C_4mim][PF_6]$ as solvents for the transition metal catalysed oligomerization of ethylene to higher olefins.[3] They carried out the biphasic process using heptane as the organic phase and an ionic liquid phase containing a cationic nickel complex as catalyst. The process gave predominantly linear alk-1-ene products with better reactivity and selectivity than in conventional organic solvents such as dichloromethane, CH_2Cl_2. After the reaction, the products separated as a clear and colourless organic layer from the ionic liquid–catalyst solution. The solution was then recycled with little change in selectivity. The group was unable to detect any leaching of the catalyst into the organic layer.

Evidence over recent years, however, suggests that ionic liquids that were originally thought to act as innocent solvents may not be so innocent after all. For example, imidazolium cations have been shown to react with transition metal complexes to form *N*-heterocyclic carbene complexes.[4]

The functioning of ionic liquids as both solvent and catalyst is classically illustrated by the use of chloroaluminate ionic liquids as solvents and catalysts for Friedel–Crafts alkylations and acylations (Section 11.5). In 1986, for example, Wilkes and co-workers demonstrated that a binary mixture of [C$_2$mim]Cl and AlCl$_3$ may be used to alkylate and acylate benzene.[5] The chemists used an excess of AlCl$_3$ in their binary mixture to ensure that the Lewis acid catalyst [Al$_2$Cl$_7$]$^-$ was present in large excess.

For the alkylations, the team added a series of primary and secondary alkyl halides to mixtures of the ionic liquid and benzene. They obtained monoalkylated and polyalkylated benzenes. The team also used the ionic liquid as solvent and catalyst for the acylation of benzene to form acetophenone, using acetyl chloride as the acylating agent.

Ionic liquids that exhibit Lewis acidity can also be exploited as solvents and co-catalysts for transition metal catalysis. The chlorogermanate and chlorostannate ionic liquids used by Parshall for platinum-catalysed reactions in the early 1970s are examples.[1]

Typically, a Lewis acidic ionic liquid interacts with a catalyst precursor to generate a cationic complex that is catalytically active. Wassercheid and Waffenschmidt have shown, for example, that chlorostannate ionic liquids activate a platinum catalyst used for the regioselective hydroformylation of 1-octene, C$_6$H$_{13}$CH=CH$_2$.[6] Hydroformylation is the reaction of carbon monoxide, hydrogen and an alkene to form linear and branched aldehydes.

The two chemists prepared two slightly Lewis acidic chlorostannate ionic liquids containing [SnCl$_3$]$^-$ and acidic [Sn$_2$Cl$_5$]$^-$ anions by mixing [C$_4$mim]Cl and 1-butyl-4-methylpyridinium chloride with a slight excess of SnCl$_2$. The acidic anions interact with the platinum catalyst, [P(C$_6$H$_5$)$_3$]$_2$PtCl$_2$, to remove chlorine atoms and form the catalytically active cationic species [(P(C$_6$H$_5$)$_3$)$_2$PtCl]$^+$.

Wassercheid and Waffenschmidt compared the platinum-catalysed hydroformylation of 1-octene in the two ionic liquids with that in CH$_2$Cl$_2$, a conventional organic solvent. The platinum catalyst showed enhanced stability and selectivity for the linear aldehyde over the branched aldehyde in both ionic liquids compared with that in CH$_2$Cl$_2$. In addition, whereas the reaction in CH$_2$Cl$_2$ is monophasic, the chlorostannate ionic liquid reactions are biphasic. The ionic liquid–catalyst solutions can therefore be recovered by phase separation and reused.

Ionic liquids incorporating transition metal containing anions or cations have also been developed. In 2003, for example, Dyson, McIndoe and Zhao described the preparation of an ionic liquid with a transition metal carbonyl anion.[7] The team prepared [C$_4$mim][Rh(CO)$_2$I$_2$] by the

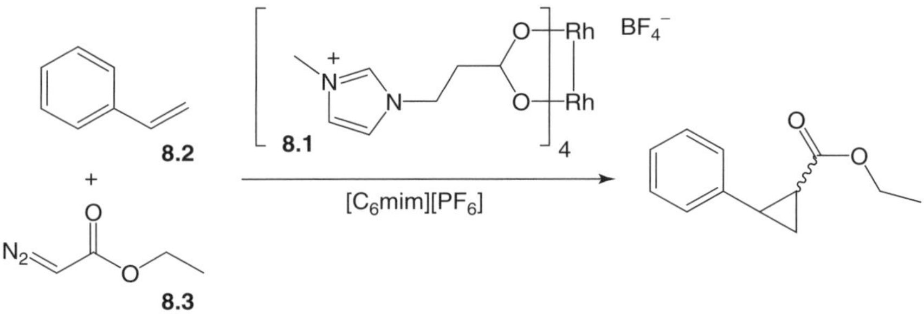

Scheme 8.1 Cyclopropanation catalysed by an ionic liquid-rhodium complex.

reaction of [C$_4$mim]I and [Rh$_2$(CO)$_4$I$_2$]. [Rh(CO)$_2$I$_2$]$^-$ is known as the Monsanto catalyst anion as it is the catalytically active species in the Monsanto process for the manufacture of acetic acid from methanol and carbon monoxide. Dyson's group used electrospray ionization ion trap mass spectrometry to analyse the catalyst in its ionic liquid form (Section 13.9).

An example of an ionic liquid catalyst with a transition-metal-containing cation was described by Forbes and co-workers in 2006.[8] The chemists prepared an ionic liquid consisting of a [BF$_4$]$^-$ anion and an imididazolium cation with a rhodium(II) complex covalently tethered to its ring (Scheme 8.1). The imidazolium dirhodium(II) carboxylate cations **8.1** were prepared by exchanging the acetate ligands of dirhodium(II) acetate, Rh$_2$(OAc)$_4$, with an imidazolium cation functionalized with carboxylic acid. The ionic liquid rhodium conjugate was found to be an effective catalyst for the intermolecular cyclopropanation reaction of styrene (**8.2**) and ethyl diazoacetate (**8.3**). The ionic liquid [C$_6$mim][PF$_6$] was used as a solvent for the reaction.

Cations of dialkylimidazolium ionic liquids have also been used to generate transition-metal-carbene complexes. Dialkylimidazolium cations have an acidic hydrogen atom on the carbon atom at the 2-position of the imidazolium ring, *i.e.* the carbon atom between the two nitrogen atoms. The metal-carbene complexes are formed by deprotonation of this ring at this position by basic ligands of a metal precursor.

An early example of such a complex was described by Xiao and colleagues in 2000.[9] The group prepared a palladium imidazolylidene complex by heating palladium(II) acetate, Pd(OAc)$_2$, in [C$_4$mim]Br. The resultant *N*-heterocyclic carbene complex, which exists as a mixture of four isomers, including **8.4**, was easily isolated from [C$_4$mim]Br and characterized. In contrast, the carbene complex was not detected after heating palladium(II) acetate in [C$_4$mim][BF$_4$] under the same conditions.

The formation of the complex in [C$_4$mim]Br can be explained by the stronger basicity of the bromide anions compared with the [BF$_4$]$^-$ anion. Xiao's team showed that the complex exhibited higher catalytic activity and stability in Heck C–C coupling reactions (Section 11.7) in [C$_4$mim]Br than in [C$_4$mim][BF$_4$].

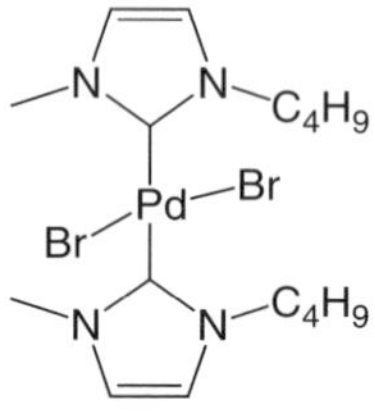

8.4

Several papers have shown that grafting ionic-liquid-type cations onto ruthenium and rhodium catalysts enhances the solubility of the catalysts in ionic liquids and allows their reuse without loss of catalytic activity. In one such paper, Lee and co-workers reported attaching two 1,2-dimethylimidazolium tags to the 1,4-bisphosphine ligand of a chiral rhodium complex with 1,5-cyclooctadiene (cod) ligands.[10] The resulting dicationic bisphosphine ligand **8.5** not only avoids catalyst leaching but also increases the stability of the catalyst in the ionic liquid [C$_4$mim][SbF$_6$].

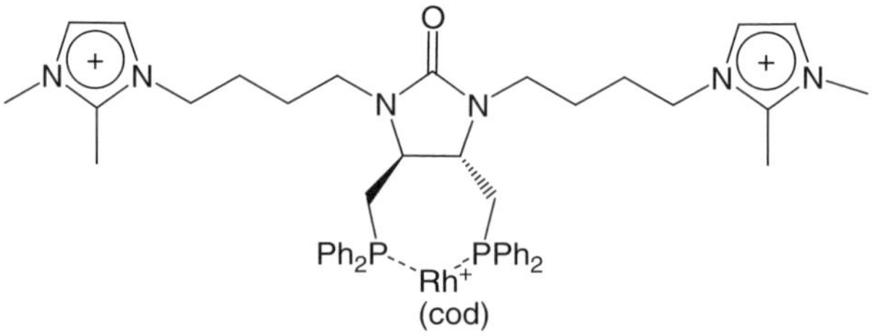

8.5

Lee's group employed a biphasic system consisting of [C$_4$mim][SbF$_6$] as one phase and iso-propanol (*i*PrOH) as the other. The catalyst, immobilized in the ionic liquid, was reused several times for the asymmetric hydrogenation of *N*-acetylphenylethenamine in over 95% ee (enantiomeric excess) without loss of activity (Scheme 8.2).

In another example of the immobilization of a catalyst in an ionic liquid, Wasserscheid and colleagues used an ammoniumphosphine ligand to immobilize and stabilize a palladium catalyst in [C$_4$mim][PF$_4$].[11] They used the catalyst to carry out the ionic liquid–organic solvent

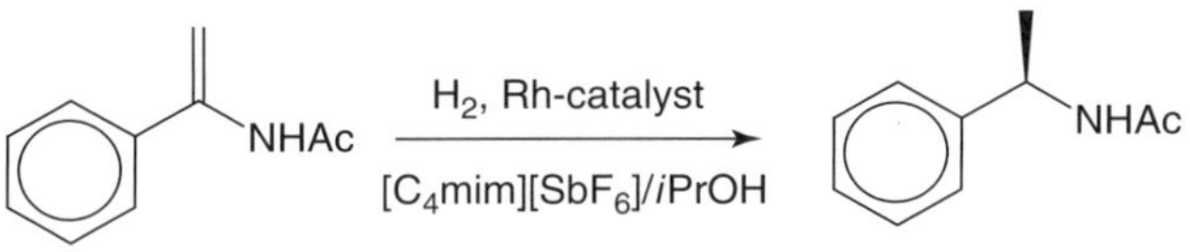

Scheme 8.2 Rhodium-catalysed asymmetric hydrogenation.

biphasic dimerization of methyl acrylate in a continuous flow apparatus. Toluene was used as the organic phase to introduce the methyl methacrylate and remove the reaction product, dihydrodimethylmuconate.

Continuous flow homogeneous catalysis has also been carried using biphasic mixtures of an ionic liquid–catalyst solution and a supercritical CO_2 phase. The ionic liquid solution acts as a stationary reactive phase. The supercritical CO_2 phase transports the substrate and reactants to the catalyst and extracts the products from the ionic liquid phase.

Cole-Hamilton and colleagues, for example, used a mobile supercritical CO_2 phase and rhodium complexes dissolved in [C_4mim][PF_6] as the stationary phase for the hydroformylation of alkenes in a continuous flow process.[12] A mixture of supercritical CO_2, carbon monoxide, hydrogen and the alkene was passed through a reactor containing a solution of the rhodium catalyst in the ionic liquid. Careful selection of ligands for the complex prevented degradation of the catalyst and rhodium leaching into the product.

8.4 CATALYST PERFORMANCE

The performance of a catalyst is measured in various ways. The number of moles of a substrate that are converted into a product by a mole of the catalyst before the catalyst becomes deactivated is one measure of the activity of a catalyst. That measure is known as the turnover number. Turnover frequency, which is the turnover per unit time, is another measure of catalytic activity. Catalyst performance also relates to the yield of a desired product and the selectivity for that product over other products of the reaction.

The ability of ionic liquids to enhance the performance of catalysts used in homogeneous catalysis, particularly in liquid–liquid biphasic systems, has attracted the attentions of chemists around the world and led to studies of a wide variety of catalytic reactions in ionic liquids.

In one such investigation, Song and co-workers showed that ionic liquids containing non-coordinating anions can "dramatically" enhance the catalytic performance of the metal triflate-catalysed Friedel–Crafts alkenylation of arenes with alkynes.[13] For example, for the alkenylation

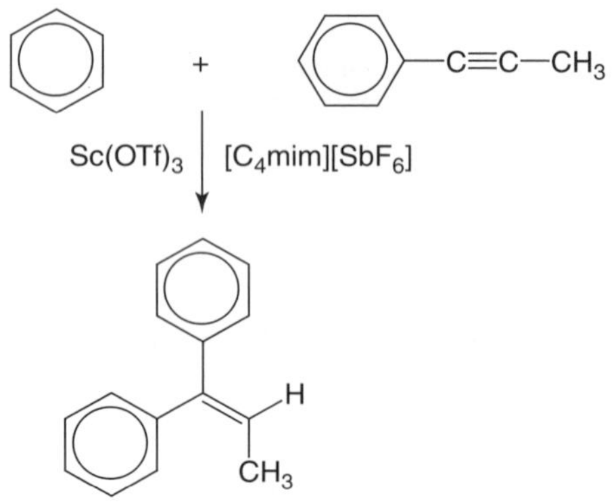

Scheme 8.3   Alkenylation of benzene with 1-phenyl-1-propyne.

of benzene with 1-phenyl-1-propyne, the team achieved a 90% yield of the product, 1,1-diphenyl-1-propene, using the catalyst scandium triflate, $Sc(OTf)_3$, and the ionic liquid $[C_4mim][SbF_6]$ (Scheme 8.3). Without the ionic liquid, the yield was 27%.

As the process is biphasic, the catalyst-containing ionic liquid phase can be easily separated after the reaction by decantation of the organic layer. The chemists demonstrated that the ionic liquid–catalyst phase could be reused repeatedly without any significant loss of activity. They also showed that the metal-triflate catalysed Friedel–Crafts alkenylations in ionic liquids could be used for a broad range of arenes and alkynes. In some cases, such reactions are not possible in conventional organic solvents.

In another example, Dyson and co-workers compared the catalytic activity of a ruthenium phosphine catalyst for the hydrogenation of arenes in $[C_4mim][BF_4]$ with its activity in dichloromethane, CH_2Cl_2.[14] They showed that the complex is "considerably more" active in the ionic liquid. The catalyst is also highly selective in the ionic liquid. For example, the catalyst hydrogenated the arene ring of allylbenzene but not the alkene bond in the allyl group ($CH_2=CHCH_2-$). The turnover frequency for the reaction in $[C_4mim][BF_4]$ was 329 moles of the product, allylcyclohexane, per mole of catalyst per hour. The turnover frequencies for the hydrogenation of toluene and ethylbenzene in the ionic liquid were 205 and $127\,mol\,mol^{-1}\,h^{-1}$, respectively. These turnover frequencies are "significantly" higher than those for the hydrogenations using the same catalyst in dichloromethane. Furthermore, the catalyst decomposed in dichloromethane over a number of runs whereas the team observed no decrease in catalytic activity in the ionic liquid after five runs.

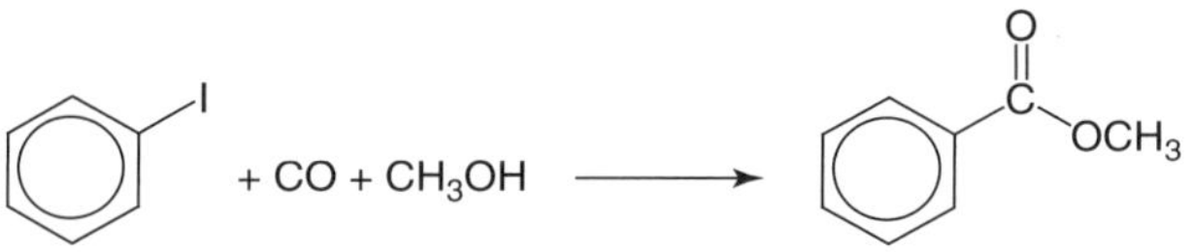

Scheme 8.4 Methoxycarbonylation of iodobenzene.

The performance of catalysts in ionic liquids varies immensely, depending on the nature of the cations and the anions. This variation is illustrated by studies carried out by Trzeciak and co-workers on the palladium-catalysed methoxycarbonylation of iodobenzene in a range of ionic liquids.[15] In this process, methanol and carbon monoxide react with iodobenzene to form methyl benzoate (Scheme 8.4).

Trzeciak's group dissolved the catalyst, a palladium(0) colloid stabilized with a polymer, in several pyridinium- and imidazolium-based ionic liquids. They found that the highest yields, ranging from 60 to 83% of the benzoic acid methyl ester, were obtained in the pyridinium ionic liquids. The lowest yield, just 8%, was obtained when [C$_4$mim]Cl was used as a solvent. The group concluded that all the imidazolium ionic liquids, with the exception of [C$_4$mim][PF$_6$], inhibited the methoxy-carbonylation reaction.

Trzeciak and co-workers subsequently carried out a more detailed study of the inhibiting effect of imidazolium halides on the palladium-catalysed methoxycarbonylation of iodobenzene.[16] This time, they used various palladium(II) complexes as catalysts. The use of [C$_4$mim][BF$_4$] and [C$_4$mim][PF$_6$] as reaction media resulted in yields of 50–78% of the ester, whereas in [C$_4$mim]Br and [C$_4$mim]Cl the reaction was totally retarded. However, when the C-2 proton on the imidazolium ring of the halide ionic liquids was replaced with a methyl group (**8.6**), the inhibiting effect was not observed. The authors suggested that imidazolium halide liquids with cations containing C-2 protons form a halide-containing palladium-carbene complex that inhibits the methoxy-carbonylation.

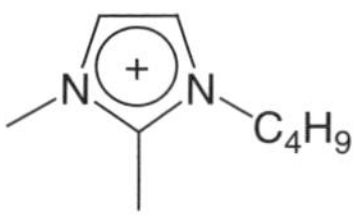

8.6

Catalysts can also be deactivated by impurities in ionic liquids. Stark and co-workers, for example, showed that transition metal catalysts in ionic liquids are "extremely" sensitive to three impurities: halides, water

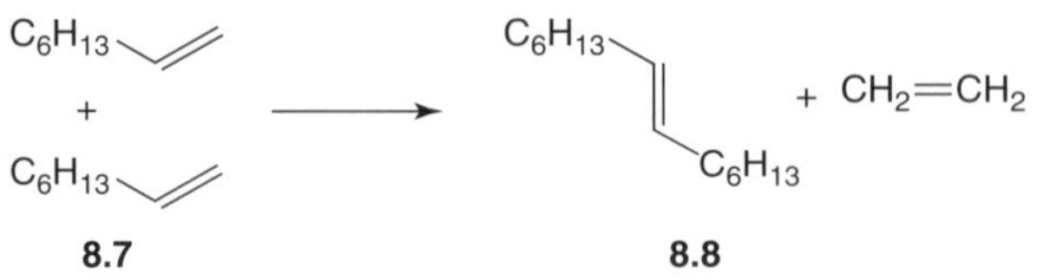

C_6H_{13}

+

C_6H_{13}

8.7

$\longrightarrow$

C_6H_{13}

C_6H_{13}

8.8

$+ \; CH_2{=}CH_2$

Scheme 8.5   Metathesis of 1-octene to 7-tetradecene.

and imines such as 1-methylimidazole.[17] The team reached their conclusions by studying the metathesis of 1-octene (**8.7**) to 7-tetradecene
(**8.8**) using various ruthenium-based precursors in imidazolium- and
pyridinium-based ionic liquids with [BF$_4$]$^-$ anions (Scheme 8.5).

8.5 SUPPORTED IONIC LIQUID CATALYSIS

Much of the research on catalysis in ionic liquids has focused on their
use as homogeneous catalysts, particularly in liquid–liquid biphasic
systems that allow the ionic liquid–catalyst solutions to be readily
separated from the product-containing organic phase. In the chemical
industry, however, heterogeneous catalysts are frequently used, particularly in continuous flow processes. In such processes, the reactants
typically flow over a fixed-bed reactor containing a solid catalyst. The
products are therefore readily separated from the catalyst.

Two related techniques known as supported ionic liquid catalysis
(SILC) and supported ionic liquid phase (SILP) catalysis combine the
benefits of homogeneous catalysis, such as high specificity and selectivity, with those of heterogeneous catalysis, particularly ease of product
separation. The reduced usage of ionic liquids in both techniques and
the potential use of these systems for fixed-bed technology are additional
potential benefits. Silica, SiO$_2$, is typically used as support in SILC and
SILP catalysis.

The terms SILC and SILP are sometimes used interchangeably. With
SILC, the ionic liquid acts as the catalyst. It is immobilized on the
support by covalent attachment of the ionic liquid to the surface of the
support. Physisorption of the ionic liquid on the support or electrostatic
interaction with the support are also used to immobilize the solution.

In SILP catalysis, a thin film of a solution of an ionic liquid and a
homogeneous catalyst is employed. The film serves as a reaction phase.
SILP catalysis has been used for a wide range of transition metal
catalysed reactions. They include rhodium-catalysed hydrogenations,
hydroformylations and carbonylations, and palladium-catalysed C–C
coupling reactions.

An early example of SILP catalysis, described by Mehnert and co-workers in 2002, focused on the use of a rhodium phosphine catalyst dissolved in a layer of [C₄mim][PF₆] for the hydrogenations of hexene-1, cyclohexene and 2,3-dimethyl-2-butene.[18] Silica gel was used as a support and heptane as the "run solvent" that carried the substrates over the supported catalyst. The catalyst was reused over 18 batch runs without loss of activity. It also exhibited enhanced activity in comparison with hydrogenations carried out using either a homogeneous system consisting of an acetone solution of the same catalyst or liquid–liquid biphasic catalysis with an ionic liquid phase.

SILP catalysis has also been carried out in continuous flow rather than batch mode. Cole-Hamilton and colleagues employed a fixed-bed SILP catalyst in a tubular reactor for the continuous flow hydroformylation of 1-octene. They used supercritical CO_2 to transport the 1-octene and reacting gases, carbon monoxide and hydrogen, over the fixed bed.[19] The bed consisted of a rhodium catalyst dissolved in [C₈mim][NTf₂] supported by silica gel. The supercritical fluid not only separated the reaction products, linear and branched aldehydes, from the catalyst but also removed traces of heavier aldol condensation products that might otherwise have fouled the catalyst. The team showed that the hydroformylation could be continuously carried out for more than 40 hours without loss of catalytic activity.

In 2002, Hölderich and co-workers described several methods for immobilizing Lewis acidic ionic liquids for SILC.[20] Typically, the cations of an imidazolium chloride were covalently bound to the silicon atoms on the surface of a silica support material (Figure 8.1). A metal halide, such as $AlCl_3$, was then added. Hölderich's team observed that the catalysts exhibited catalytic activities for Friedel–Crafts alkylations of benzene with different alkenes that were "among the best to be found in the literature".

Supports other than SiO_2 have also been tested for SILC and SILP. For example, Serp and co-workers used high purity multiwalled carbon nanotubes as a support for a rhodium phosphine catalyst immobilized in a film of [C₄mim][PF₆].[21] The surfaces of the nanotubes were functionalized

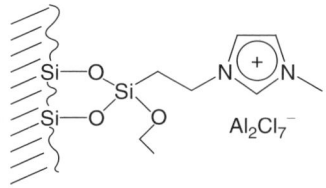

Figure 8.1   ionic liquid cation covalently tethered to a support.

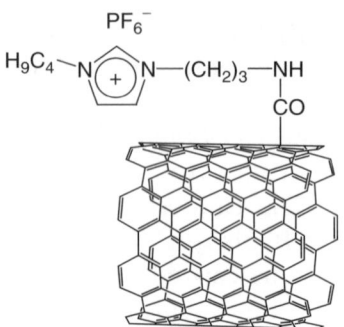

Figure 8.2 Ionic liquid immobilized on a carbon nanotube.

with imidazolium-based ionic moieties (Figure 8.2) to avoid leaching of the ionic liquid film and allow recycling of the catalyst system.

The group evaluated the catalytic activity of the supported catalyst for the hydrogenation of 1-hexene to *n*-hexane in the solvent heptane. The researchers showed that the nanotube-supported catalyst system exhibits high stability and higher catalytic activity for the reaction than the same SILP catalyst system on silica and other oxide supports. They observed selectivity of up to 100% for the *n*-hexane product when the catalyst system was recycled.

8.6 NANOPARTICLE CATALYSTS

By the end of the 1990s, transition metal nanoparticle catalysts had been extensively studied for a wide range of reactions in conventional organic solvents.[22] Compared with their bulk heterogeneous counterparts, nanoparticle catalysts offer several potential advantages. In particular, the large surface areas of the nanoparticles allow larger numbers of catalytically active metal centres to be exposed to the reactants.

Ionic liquids have proved to be suitable solvents for the preparation and stabilization of nanoparticle catalysts. One of the earliest papers on the topic, published in 2001,[23] focused on the preparation and stabilization of palladium nanoparticles, Pd(0), in ionic liquids and their use for Heck C–C coupling reactions (Section 11.7).

The chemists carried out the reactions in two imidazolium-based ionic liquids containing Pd(OAc)$_2$ or PdCl$_2$. The reaction mixture was subjected to ultrasound irradiation, leading to the formation of catalyst precursors: palladium-biscarbene complexes such as **8.4**. The precursors subsequently reduced to the active catalytic species: nanoparticles of Pd(0). These nanoparticles were highly stable and considerably enhanced the rate of reaction compared with Pd-catalysed reactions in ionic liquids without the use of ultrasonic irradiation.

The following year, Dupont's team described the preparation and stabilization of iridium nanoparticles, Ir(0), in ionic liquids and showed that they could be used as recyclable catalysts for biphasic hydrogenation reactions.[24] The nanoparticles were formed by the reduction of an Ir(I) precursor with 1,5-cyclooctadiene (cod) ligands, [IrCl(cod)]$_2$, in [C$_4$mim][PF$_6$]. The catalytic activity of the system was "significantly superior" to those obtained when ionic liquids and classical transition-metal precursors are used for biphasic hydrogenations under similar reaction conditions.

Finke and co-workers used ^{2}H-nuclear magnetic resonance spectroscopy to show that Ir(0) nanoclusters react with imidazolium-based ionic liquids to generate N-heterocyclic carbenes attached to the nanocluster surfaces.[4] The results suggest that it is the surface-attached carbenes that stabilize the nanoclusters. The results also provide evidence of the non-innocent nature of ionic liquid solvents.

The addition of ionic polymer additives to ionic liquids or the functionalization of ionic liquids can help to stabilize and improve the performance of catalyst nanoparticles in ionic liquids. Dyson and co-workers combined both approaches for stabilizing rhodium nanoparticles for the biphasic hydrogenation of styrene, C$_6$H$_5$CH=CH$_2$, to ethylbenzene, C$_6$H$_5$CH$_2$CH$_3$.[25] They showed that rhodium nanoparticles stabilized with poly(vinyl pyrrolidone) (PVP) are highly soluble in hydroxyl-functionalized ionic liquids such as [C$_2$OHmim][BF$_4$] (**8.9**) whereas the stabilized nanoparticles are almost insoluble in the non-functionalized ionic liquid [C$_2$mim][BF$_4$]. The activity, stability and recyclability of the PVP-protected rhodium nanoparticles in the functionalized ionic liquids proved to be superior to that in non-functionalized ionic liquids.

8.9

Nanoparticles can also play a role as supports for transition metal catalysts in ionic liquids. One example is the use of carbon nanotubes as a support for a rhodium phosphine catalyst immobilized in an ionic liquid (Section 8.5).

In another example, Yu and colleagues demonstrated that molecular Pd(II) complex catalysts immobilized on gold nanoparticles are highly effective catalysts in the ionic liquid [C$_4$mim][PF$_6$] for the microwave-assisted cyclotrimerization of alkynes, such as $CH_3C\equiv$ CC_2H_5.[26] The products of this reaction are benzene rings with alkyl substituents on each of the six carbon atoms. The catalysts were readily separated from the reaction mixtures by centrifugation and filtration and reused many times without loss of catalytic activity.

8.7 ACID–BASE CATALYSIS

Acidic chloroaluminate ionic liquids can act as solvents and their $[Al_2Cl_7]^-$ anions as Lewis acid catalysts for Friedel–Crafts alkylations and acylations (Sections 8.3 and 11.5).

Ionic liquids have also been used as solvents for Brønsted acid catalysed reactions. An early example, reported by Raston and co-workers in 2000, was the acid-catalysed synthesis of cyclotriveratrylene (CTV, **8.10**) using the ionic liquid $[N_{6\,4\,4\,4}][NTf_2]$ as a reaction medium.[27] CTV is a supramolecular host compound that complexes with various guest molecules ranging from volatile solvent molecules to the fullerene C_{60}.

The team synthesized the compound by cyclotrimerization of veratryl alcohol (**8.11**, Scheme 8.6). The condensation process employed phosphoric acid, H_3PO_4, as a catalyst and was carried out under mild conditions. The conventional synthesis of CTV and related compounds requires large volumes of volatile organic solvents and strongly acidic and dehydrating conditions. The white crystalline product was obtained in high yield and purity and the ionic liquid readily recovered for reuse.

There have also been several studies of base catalysis in ionic liquids. In 1998, for example, Earle and co-workers published a paper on two base-catalysed regioselective alkylations in the ionic liquids $[C_4mim][PF_6]$ and $[C_4mim][BF_4]$.[28]

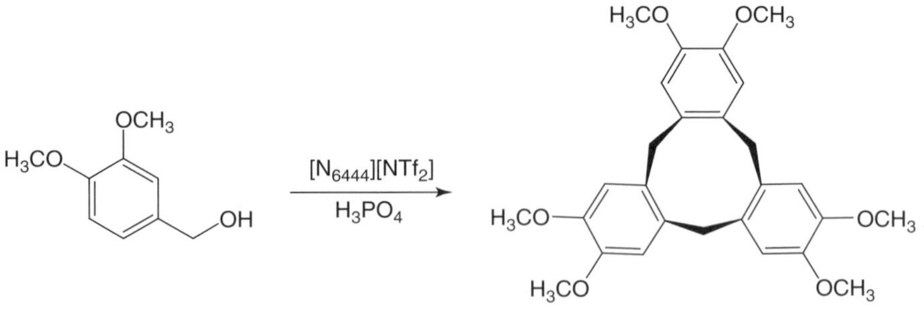

8.11   **8.10**

Scheme 8.6 Synthesis of cyclotriveratrylene.

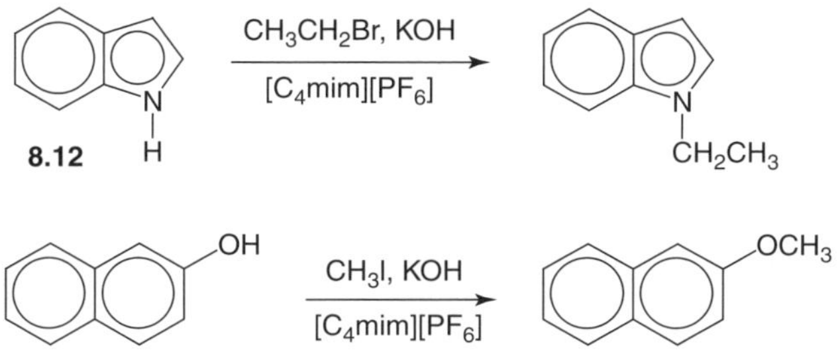

Scheme 8.7   Base-catalysed regioselective alkylations.

Regioselective alkylations are solvent-dependent nucleophilic displacement reactions that are normally carried out in polar aprotic solvents, such as dimethyl sulfoxide (DMSO) or dimethylformamide (DMF). These solvents accelerate the reactions. However, the relatively high boiling points of DMSO and DMF, their thermal instability, noxious and toxic odours, and miscibility with aqueous and organic solvents are major drawbacks. As a result, separation of high purity products is difficult and recovery of the solvents is virtually impossible. [C$_4$mim][PF$_6$] and [C$_4$mim][BF$_4$], in contrast, give similar rate enhancements to DMSO and DMF, enable the products to be readily isolated, and are readily recycled.

Earle and colleagues tested the two ionic liquids as solvents for the alkylation of the nitrogen atom of indole (**8.12**) and the oxygen atom of 2-naphthol (**8.13**, Scheme 8.7). They carried out the reactions at room temperature using simple halides, such as ethyl bromide and methyl iodide, as alkylating agents and potassium hydroxide as the base. The products were extracted from the reaction mixture with diethyl ether in high yields.

Forsyth, MacFarlane and co-workers have described the use of Lewis base ionic liquids as solvents and catalysts for the base-catalysed acetylation of alcohols and carbohydrates.[29,30] They showed that a range of alcohols and carbohydrates can be rapidly and fully acetylated at room temperature by acetic anhydride in ionic liquids such as [C$_4$mim][dca], which contain the basic dicyanamide anion. The acetylated products were precipitated from the ionic liquid solutions by addition of water. The reactions in ionic liquids were not only faster but also higher yielding than in pyridine which is conventionally employed as a solvent–catalyst for this process.

8.8   TRANSITION METAL CATALYSIS

Interest in transition metal catalysis in ionic liquids can be traced back to the work of Parshall, published in 1972, on platinum-catalysed reactions in tetraethylammonium trichlorogermanate and trichloro-stannate ionic liquids (Section 8.2).[1]

Chauvin's group also carried out some of the early investigations of the use of ionic liquids for transition metal catalysis. Chauvin shared the Nobel Prize in Chemistry in 2005 for "the development of the metathesis method in organic synthesis". He was particularly interested in alkene metathesis – a reaction in which carbon–carbon double bonds in alkenes relocate to form different olefins (Scheme 8.5).

In 1990, his group published a paper on the catalytic dimerization of propene to hexenes by nickel complexes using room-temperature "organochloroaluminate molten salts" as solvents.[31] The salts were based on mixtures of $AlCl_3$ or ethylaluminium dichloride, $Al(C_2H_5)Cl_2$, and $[C_4mim]Cl$, $[C_4py]Cl$ or $[P_{4\ 4\ 4\ 4}]Cl$. Chauvin's team observed catalytic activity in acidic melts of these ionic liquids but not in the basic melts. Until then, no attempt had been made "to take advantage of the solubility of an organometallic catalyst and the insolubility of the re-action products of the catalytic reaction in these solvents", the authors noted.

In 1995, Chauvin's team reported the use of "a novel class of versatile solvents for two-phase catalysis".[32] The group showed that rhodium complexes with water-soluble phosphine ligands dissolved in room-temperature 1,3-dialkylimidazolium ionic liquids with non-nucleophilic anions, such as $[BF_4]^-$ and $[PF_6]^-$, catalyse the hydrogenation, iso-merization and hydroformylation of pentene. The ionic liquid phase was recycled after removal of the product-containing organic phase.

Chauvin and Olivier-Bourbigou also obtained "good results" for the metathesis of 2-pentene, $CH_3CH=CHCH_2CH_3$, using a tungsten cata-lyst dissolved in ionic liquids.[33] The products of the reaction are 2-butene, $CH_3CH=CHCH_3$, and 3-hexene, $CH_3CH_2CH=CHCH_2CH_3$. The two chemists noted, once again, that the catalyst remains in the ionic liquid phase and can be reused several times after decantation of the hydrocarbon layer. Ionic liquids show great promise for two-phase catalysis, they remarked in their paper. They termed the solvents "nonaqueous ionic liquids", or NAILs, and observed that their chemical properties such as complexing ability and acidity can be tuned at will.

In 1996, De Souza, Dupont and co-workers described the use of "new" ionic liquids for rhodium-catalysed hydrogenations.[34] They demonstrated that rhodium complexes are completely soluble in $[C_4mim][BF_4]$ and $[C_4mim][PF_6]$ and that they catalyse the hydrogen-ation of cyclohexene in a two-phase process. The products were removed by simple decantation, leaving the rhodium catalysts "almost com-pletely" retained in the ionic liquid.

Over the past ten years or so, there have been countless studies of the use of transition metal catalysts in ionic liquids for hydrogenations, oxidations, hydroformylations, C–C coupling reactions, dimerizations, oligomerizations, olefin metathesis and other types of reaction. Further examples of these reactions are described in Chapters 10 and 11.

8.9 CHIRAL TRANSITION METAL CATALYSIS

Chiral transition metal catalysis is widely used for the synthesis of enantiomerically-enriched compounds. In most cases, the ligands of the metal complexes provide the chiral information.

The use of such chiral complexes for asymmetric synthesis in achiral ionic liquids has been studied extensively in recent years. In addition, the development of chiral ionic liquids as solvents or additives for asymmetric catalysis over recent years has opened up new possibilities for enantioselective transformations.

One of the earliest examples of the use of chiral catalysts in ionic liquids for asymmetric synthesis, described by Chauvin's group in 1995, used the biphasic [C_4mim][SbF_6]/isopropanol system.[32] The group showed that the enantioselective hydrogenation of α-acetamidocinnamic acid (**8.14**) using a chiral cationic rhodium complex yielded (*S*)-*N*-acetylphenylalanine (**8.15**) in 64% ee. The product was separated quantitatively and the ionic liquid recycled.

$$\text{CO}_2\text{H}$$
$$\text{NHCOCH}_3$$

8.14

$$\text{CO}_2\text{H}$$
$$\text{NHCOCH}_3$$

8.15

Following the early work by Chauvin and co-workers, interest in asymmetric synthesis using chiral catalysts in ionic liquids began to grow, most notably around 2000. In 2000, for example, Song and Roh employed [C_4mim][PF_6] to immobilize, recover and recycle a homogeneous chiral catalyst that is used for the asymmetric epoxidation of alkenes.[35] The two scientists immobilized a catalyst, known as Jacobsen's chiral manganese(III) salen epoxidation catalyst (**8.16**), in the ionic liquid. They tested the immobilized catalyst for the asymmetric epoxidation of several alkenes in a [C_4mim][PF_6]/CH_2Cl_2 biphasic system. For example, for the epoxidation of

2,2-dimethylchromene (**8.17**) they achieved 96% ee for the (*R,R*)-product **8.18** (Scheme 8.8). They demonstrated that the immobilized catalyst can be recycled five times with only a slight decrease in enantioselectivity.

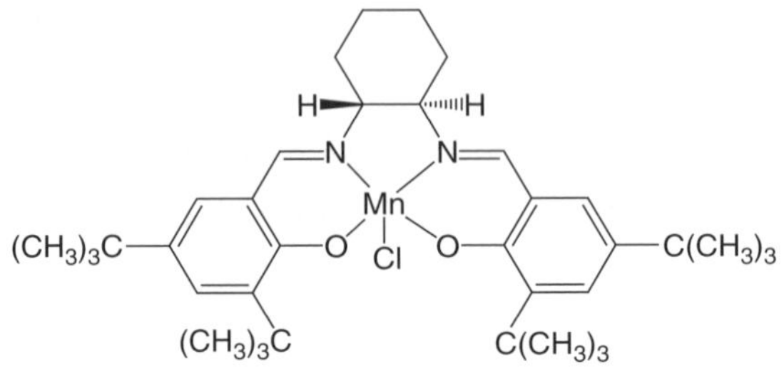

8.16

In the same year, Song's group described the chromium(III) salen catalysed asymmetric ring-opening reactions of cyclopentene oxide with trimethylsilyl azide, $(CH_3)_3SiN_3$, in several ionic liquids, using a biphasic system with hexane as the upper phase.[36] They showed that the catalytic activity and enantioselectivity depended strongly on the nature of the ionic liquid anion. For example, they achieved "excellent results" for the hydrophobic ionic liquid [C$_4$mim][PF$_4$] but little or no reaction in the hydrophilic ionic liquids [C$_4$mim][BF$_4$] or [C$_4$mim][OTf].

By 2003 numerous examples of chiral transition metal catalysis in achiral ionic liquids had been published.[37] Within a few years, an increasing range of chiral ionic liquids were also being developed for enantioselective catalysis.

In 2007, for example, Franciò and co-workers described the use of ionic liquids with chiral cations for enantioselective homogeneous rhodium-catalysed hydrogenations.[38] Chiral ionic liquids such as **8.19** were prepared from amino acids such as L-proline. The catalyst was formed by dissolving "tropos" biphenylphosphine ligands **8.20** modified with sulfonato groups in the ionic liquid. The phenyl rings in these ligands rotate rapidly and are therefore not chiral. They become chiral in the presence of an external chiral bias which, in this case, was provided by the chiral cations of the ionic liquid. The sulfonato groups help the ligand to dissolve and become

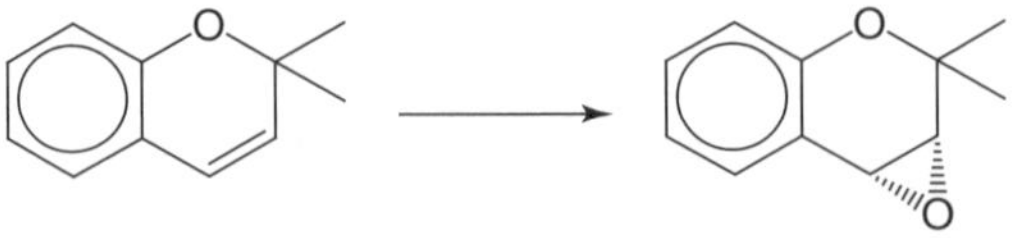

8.17   **8.18**

Scheme 8.8   Asymmetric epoxidation of 2,2-dimethylchromene.

immobilized in the ionic liquid. A solution of the rhodium precursor [Rh(cod)$_2$][BF$_4$] (where cod=1,5-cyclooctadiene) in CH$_2$Cl$_2$ was finally added to the chiral ionic liquid–ligand solution.

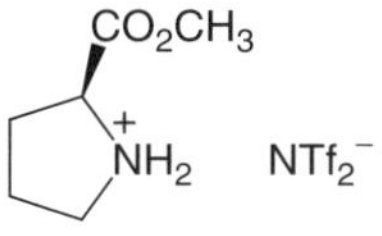

8.19 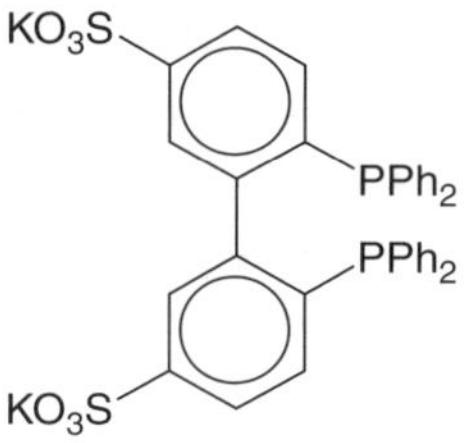   **8.20**

The team obtained "significant enantioselectivities" using the chiral ionic liquid–catalyst system for the hydrogenation of methyl 2-acetamidoacrylate (**8.21a**) and dimethyl itaconate (**8.21b**). Furthermore, the products could be removed by flushing the reaction mixture with supercritical CO$_2$, allowing the ionic liquid–catalyst system to be reused.

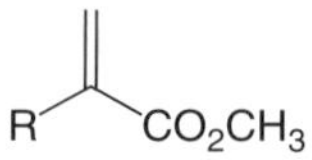

a   R = NH(CO)CH$_3$
b   R = CH$_2$CO$_2$CH$_3$

8.21

In subsequent work, the same team showed that the proline-derived chiral ionic liquid induced high levels of enantioselectivity for the hydrogenation of the same two substrates when used with a rhodium catalyst with racemic binap ligands (where binap=2,2'-bis(diphenylphosphino)-1,1'-binaphthyl, **8.22**).[39]

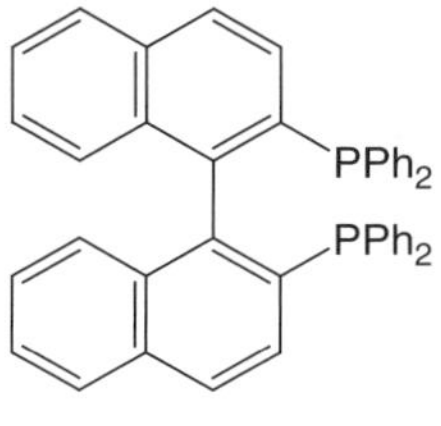

8.22

8.10   TASK-SPECIFIC IONIC LIQUIDS IN CATALYSIS

The carboxylate-functionalized ionic liquid used to prepare the imidazolium dirhodium(II) catalyst for the cyclopropanation reaction of styrene and ethyl diazoacetate (Section 8.3) and the chiral ionic liquids used for enantioselective homogeneous rhodium-catalysed hydrogenations (Section 8.9) are examples of ionic liquids designed for a specific tasks in catalytic reactions.

Another example of a task-specific ionic liquid designed for catalysis is the imidazolium ionic liquid **8.23**, which was functionalized for the ruthenium-catalysed hydrogenation of CO_2 to produce formic acid, HCOOH.[40] In this case, the imidazolium ring of the ionic liquid was functionalized with a tertiary amino group (-N(CH_3)_2) on the cation to facilitate the recovery of the formic acid and reuse of the ionic liquid and catalyst.

8.23

To date, most studies of the use of task-specific ionic liquids for catalysis have focused on functionalization of cations. Various examples of such studies are described in Chapters 10 and 11.

REFERENCES

1. G. W. Parshall, *J. Am. Chem. Soc.*, 1972, **94**, 8716.
2. P. Wasserscheid and P. Schulz, in *Ionic Liquids in Synthesis*, 2nd edn, ed. P. Wasserscheid and T. Welton, Wiley-VCH, Weinheim, 2008, **vol. 2**, p. 369.
3. P. Wasserscheid, C. M. Gordon, C. Hilgers, M. J. Muldoon and I. R. Dunkin, *Chem. Commun.*, 2001, 1186.
4. L. S. Ott, M. L. Cline, M. Deetlefs, K. R. Seddon and R. G. Finke, *J. Am. Chem. Soc.*, 2005, **127**, 5758.
5. J. A. Boon, J. A. Levisky, J. L. Pflug and J. S. Wilkes, *J. Org. Chem.*, 1986, **51**, 480.
6. P. Wasserscheid and H. Waffenschmidt, *J. Mol. Catal. A: Chem.*, 2000, **164**, 61.
7. P. J. Dyson, J. S. McIndoe and D. Zhao, *Chem. Commun.*, 2003, 508.
8. D. C. Forbes, S. A. Patrawala and K. L. T. Tran, *Organometallics*, 2006, **25**, 2693.
9. L. Xu, W. Chen and J. Xiao, *Organometallics*, 2000, **19**, 1123.

10. S. Lee, Y. J. Zhang, J. Y. Piao, H. Yoon, C. E. Song, J. H. Choi and J. Hong, *Chem. Commun.*, 2003, 2624.
11. J. Zimmermann, P. Wasserscheid, I. Tkatchenko and S. Stutzmann, *Chem. Commun.*, 2002, 760.
12. M. F. Sellin, P. B. Webb and D. J. Cole-Hamilton, *Chem. Commun.*, 2001, 781.
13. M. Y. Yoon, J. H. Kim, D. S. Choi, U. S. Shin, J. Y. Lee and C. E. Song, *Adv. Synth. Catal.*, 2007, **349**, 1725.
14. C. J. Boxwell, P. J. Dyson, D. J. Ellis and T. Welton, *J. Am. Chem. Soc.*, 2002, **124**, 9334.
15. W. Wojtków, A. J. M. Trzeciak, R. Choukroun and J. L. Pellegatta, *J. Mol. Catal. A: Chem.*, 2004, **224**, 81.
16. W. Zawartka, A. M. Trzeciak, J. J. Ziółkowski, T. Lis, Z. Ciunik and J. Pernak, *Adv. Synth. Catal.*, 2006, **348**, 1689.
17. A. Stark, M. Ajam, M. Green, H. G. Raubenheimer, A. Ranwell and B. Ondruschka, *Adv. Synth. Catal.*, 2006, **348**, 1934.
18. C. P. Mehnert, E. J. Mozeleski and R. A. Cook, *Chem. Commun.*, 2002, **1**, 3010.
19. U. Hintermair, G. Zhao, C. C. Santini, M. J. Muldoon and D. J. Cole-Hamilton, *Chem. Commun.*, 2007, 1462.
20. M. H. Valkenberg, C. deCastro and W. F. Hölderich, *Green Chem.*, 2002, **4**, 88.
21. L. Rodríquez-Pérez, E. Teuma, A. Falqui, M. Gómez and P. Serp, *Chem. Commun.*, 2008, 4201.
22. J. D. Aiken III and R. G. Finke, *J. Mol. Catal. A: Chem.*, 1999, **145**, 1.
23. R. R. Deshmukh, R. Rajagopal and K. V. Srinivasan, *Chem. Commun.*, 2001, 1544.
24. J. Dupont, G. S. Fonseca, A. P. Umpierre, P. F. P. Fichtner and S. R. Teixeira, *J. Am. Chem. Soc.*, 2002, **124**, 4228.
25. X. Yang, N. Yan, Z. Fei, R. M. Crespo-Quesada, G. Laurenczy, L. Kiwi- Minsker, Y. Kou, Y. Li and P. J. Dyson, *Inorg. Chem.*, 2008, **47**, 7444.
26. Y.-Y. Lin, S.-C. Tsai and S. J. Yu, *J. Org. Chem.*, 2008, **73**, 4920.
27. J. L. Scott, D. R. MacFarlane, C. L. Raston and C. M. Teoh, *Green Chem.*, 2000, **2**, 123.
28. M. J. Earle, P. B. McCormac and K. R. Seddon, *Chem. Commun.*, 1998, 2245.
29. S. A. Forsyth, D. R. MacFarlane, R. J. Thomson and M. von Itzstein, *Chem. Commun.*, 2002, 714.
30. D. R. MacFarlane, J. M. Pringle, K. M. Johansson, S. A. Forsyth and M. Forsyth, *Chem. Commun.*, 2006, 1905.

31. Y. Chauvin, B. Gilbert and I. Guibard, *J. Chem. Soc., Chem. Commun.*, 1990, 1715.
32. Y. Chauvin, L. Mussmann and H. Olivier, *Angew. Chem. Int. Ed. Engl.*, 1995, **34**, 2698.
33. Y. Chauvin and H. Olivier-Bourbigou, *CHEMTECH*, Sept. 1995, 26.
34. P. A. Z. Suarez, J. E. L. Dullius, S. Einloft, R. F. De Souza and J. Dupont, *Polyhedron*, 1996, **15**, 1217.
35. C. E. Song and E. J. Roh, *Chem. Commun.*, 2000, 837.
36. C. E. Song, C. R. Oh, E. J. Roh and D. J. Choo, *Chem. Commun.*, 2000, 1743.
37. C. Baudequin, J. Baudoux, J. Levillain, D. Cahard, A.-C. Gaumont and J.-C. Plaquevent, *Tetrahedron: Asymmetry*, 2003, **14**, 3081.
38. M. Schmitkamp, D. Chen, W. Leitner, J. Klankermayer and G. Franciò, *Chem. Commun.*, 2007, 4012.
39. D. Chen, M. Schmitkamp, G. Franció, J. Klankermayer and W. Leitner, *Angew. Chem. Int. Ed.*, 2008, **47**, 7339.
40. Z. Zhang, Y. Xie, W. Li, S. Hu, J. Song, T. Jiang and B. Han, *Angew. Chem. Int. Ed.*, 2008, **47**, 1127.

CHAPTER 9

Inorganic Chemistry

9.1 INTRODUCTION

The study of inorganic chemistry in room-temperature ionic liquids originated in the work of Hurley and Wier in the late 1940s. They showed that metals could be electrodeposited from fused mixtures of metal chlorides and organic salts (Section 7.6).

Over the following decades there was a trickle of papers on inorganic chemistry in ionic liquids, principally focusing on electrochemistry and electrochemical applications. However, it was not until the late 1990s and early 2000s – following the discovery of air- and water-stable room-temperature ionic liquids – that research interest in the topic accelerated.

In the past few years, the physical and chemical properties of elements and their compounds from every group and period in the periodic table have been investigated to a greater or less extent. The tentacles of inorganic chemistry in ionic liquids now reach out to not only electro-chemistry and electrochemical applications (Chapter 7) but also to catalysis (Chapter 8), organic chemistry (Chapters 10 and 11) and analytical applications (Chapter 13).

9.2 MAIN GROUP CHEMISTRY

Main group elements are elements in groups 1, 2 and 13–18 in the periodic table.

An Introduction to Ionic Liquids
By Michael Freemantle
© Michael Freemantle 2010
Published by the Royal Society of Chemistry, www.rsc.org

Considerable attention has been paid to the chemistry of the main group inorganic components of room-temperature ionic liquids. For example, the chemistry of $AlCl_3$ in chloroaluminate ionic liquids is well established and the varying coordinating abilities of inorganic ionic liquid anions such as Cl^-, $[BF_4]^-$ and $[PF_6]^-$ has been the focus of much research. Furthermore, the discovery that the $[PF_6]^-$ anion in $[C_4mim][PF_6]$ undergoes hydrolysis to give hazardous hydrogen fluoride, HF, fumes has had a major impact on the selection of ionic liquids as solvents for chemical processes.[1] The air- and water-stable $[C_4mim][NTf_2]$ has replaced $[C_4mim][PF_6]$ as the ionic liquid of choice for many applications.

The study of the physical and chemical properties of main group elements and their compounds as solutes in, rather than as components of, room-temperature ionic liquids has been wide ranging. There have been numerous studies of the solubility of gases such as O_2, N_2 and CO_2 in ionic liquids (Section 5.7), the electrochemistry of compounds such as $NaNO_3$ in ionic liquids (Section 7.2) and electrochemical applications of main group elements in ionic liquids (Chapter 7). The use of ionic liquids for the synthesis of nanoparticles of main group elements such as carbon and of compounds such as alumina, Al_2O_3, has also been investigated extensively (Sections 9.7–9.11).

Main group elements and their compounds are used widely as key reagents for various organic reactions in ionic liquids. The reactions include hydrogenations, oxidations and carbonylations using H_2, O_2 and CO, respectively (Chapters 10 and 11).

As an example, Han and colleagues prepared a task-specific ionic liquid for the production of formic acid, HCOOH, by the hydrogenation of CO_2.[2] Carbon dioxide is not only a major greenhouse gas it is also a potentially inexpensive and attractive feedstock for the manufacture of organic compounds. The conversion of CO_2 into organic compounds such as formic acid in organic solvents, water and supercritical CO_2, using homogeneous or heterogeneous catalysts, has been extensively studied over the past decade or so. Inorganic or organic bases are often added to promote the conversion into formic acid. However, recovery of the acid and recycling of the base and catalyst present problems.

Han's group synthesized a new basic ionic liquid (**9.1**), which consists of imidazolium cations with tertiary amino groups, $-N(CH_3)_2$, and $[OTf]^-$ anions. The ionic liquid allowed the facile recovery of the formic acid and the catalyst to be reused. Thermogravimetric analysis revealed that the ionic liquid is stable up to 220 °C. To carry out the reaction, the chemists dispersed a ruthenium catalyst immobilized on silica in an aqueous solution of the ionic liquid.

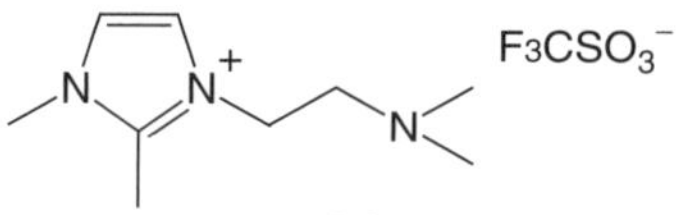

9.1

The researchers reported that the catalytic system exhibited "satisfactory activity and high selectivity." At the end of the reaction, the catalyst was removed by filtration and the water evaporated off at 110 °C. Formic acid was separated from the mixture at 130 °C in a stream of nitrogen, leaving behind the ionic liquid. The authors noted that no volatile organic substance was used and no waste produced by the process.

The chemistry of phosphorus compounds, such as phosphorus trichloride, PCl_3, and phosphorus oxychloride, $POCl_3$, in room-temperature ionic liquids has also been the subject of several investigations. PCl_3, and $POCl_3$ are employed in the organic synthesis of compounds such as dichlorophenylphosphine, $C_6H_5PCl_2$ (DCPP), which is used to manufacture pesticides, flame-retardants and other products.

The chemistry of PCl_3 in ionic liquids came into the research spotlight in 2004 when Wang and Wang carried out the Friedel–Crafts reaction of PCl_3, a liquid at room temperature, with benzene in a chloroaluminate ionic liquid to yield DCPP.[3] The ionic liquid, a binary mixture of triethylhydrogenammonium chloride and $AlCl_3$, $[(C_2H_5)_3NH]Cl$-$AlCl_3$, was used as a solvent for the reaction and also as a catalyst. The reaction mechanism for the reaction proposed by the authors is shown in Scheme 9.1.

The following year, Chen and Wang reported solubility data of PCl_3 in the same chloroaluminate ionic liquid.[4] They showed that the solubility in the ionic liquid increases with increasing temperature whereas the ionic liquid is only slightly soluble in PCl_3.

Haloaluminate ionic liquids, PCl_3 and $POCl_3$ are all sensitive to moisture and therefore require strictly anhydrous conditions to prevent their hydrolysis. In 2006, Hardacre, Migaud and colleagues demonstrated

$$PCl_3 + Al_2Cl_7^- \rightarrow Cl_2P^+ + 2AlCl_4^-$$

$$Cl_2P^+ + Ph\text{-}H \rightleftharpoons H\text{-}Ph^+ - PCl_2$$

$$H\text{-}Ph^+ - PCl_2 + AlCl_4^- \rightleftharpoons Ph\text{-}PCl_2 + HCl + AlCl_3$$

$$AlCl_3 + AlCl_4^- \rightarrow Al_2Cl_7^-$$

Scheme 9.1 Proposed mechanism for the reaction of PCl_3 and benzene in chloroaluminate ionic liquid.

that PCl_3 and $POCl_3$ exhibited unexpectedly high stability in imidazolium and pyrrolidinium ionic liquids, even when water is present.[5] The work indicated that ionic liquids are suitable not only as storage solvents for moisture-sensitive phosphorus compounds but also as reaction media for phosphorus chemistry. The group suggested that water in the ionic liquids is deactivated by hydrogen bonding and therefore does not react with the phosphorus compounds.

In 2008, the same team reported an investigation of the reactivity of PCl_3 towards primary amines, such as aniline, and secondary amines, such as diethyl amine, in three imidazolium and pyrrolidinium ionic liquids with $[NTf_2]^-$ anions.[6] The aminations lead to the formation of aminated phosphines. The team concluded that the $[NTf_2]^-$-based ionic liquids "provide a unique environment for the highly chemoselective production and storage of mono-, bis- and triaminated phosphines at $20\,^{\circ}C$" under non-anhydrous conditions.

9.3 TRANSITION METAL CHEMISTRY

Interest in transition metal chemistry in room-temperature ionic liquids has focused principally on two areas: (1) the coordination chemistry and electrochemistry of transition metal ions in haloaluminate ionic liquids and (2) organic chemistry (Chapters 10 and 11), particularly the use of ionic liquids as solvents both for transition metal catalysis (Section 8.8) and for the synthesis of organometallic compounds (Section 9.4).

Much of the early work on the coordination of transition metal ions in ionic liquids was carried out by groups led by Hussey and Osteryoung. In the late 1970s and 1980s, for example, the two groups showed in several papers that basic room-temperature chloroaluminate ionic liquids such as $[C_4py]Cl$-$AlCl_3$ and $[C_2mim]Cl$-$AlCl_3$ are "excellent" solvents for carrying out electrochemical and spectrochemical investigations of anionic transition metal chloride complexes.

These ionic liquids are basic when chloride-rich, *i.e.* when they contain an excess of the organic salt. Coordination compounds in these liquids typically consist of the organic salt cation and an anionic complex formed by the coordination of the transition metal ion with chloride ions.

In one such investigation of these systems, Osteryoung's team used absorption spectroscopy and cyclic voltammetry to study nickel(II) ion equilibria in $[C_4py]Cl$-$AlCl_3$.[7] Anhydrous $NiCl_2$ dissolves in basic $[C_4py]Cl$-$AlCl_3$ to give a sea green/blue solution. In acidic ($AlCl_3$-rich) $[C_4py]Cl$-$AlCl_3$, however, the chloride only dissolves with difficulty. In the basic ionic liquid, Ni(II) is present as $[NiCl_4]^{2-}$. In related work, Hussey's team used the same electrochemical and spectroscopic techniques to study the coordination of cobalt(II) ions in basic

[C$_4$py]Cl-AlCl$_3$.[8] The team showed that Co(II) is tetrahedrally coordinated as the [CoCl$_4$]$^{2-}$ ion in the ionic liquid.

The presence of even minute traces of water in chloroaluminate ionic liquids affords aluminium oxo- and hydroxo-containing impurities. These oxide-containing species react with transition metal ions to form transition metal oxochloride species. In 1987, Hussey's group showed that the addition of phosgene, COCl$_2$, to basic chloroaluminate ionic liquids with transition metal solutes removes all traces of transition metal oxochloride impurities.[9] The impurities are converted into chloride complexes.

By 1988, numerous transition metal chlorides had been shown to form well-defined anionic complexes in basic haloaluminate ionic liquids.[10] Haloaluminate ionic liquids have also proved to be excellent solvents for stabilizing and investigating the chemistry of transition metal halide cluster complexes. These complexes are polynuclear species in which two or more metal atoms are bound together.

In one of study of these complexes, reported in 1990, Hussey's group described the characterization of polynuclear rhenium(III) chloride cluster complexes in basic [C$_2$mim]Cl-AlCl$_3$ using absorption spectroscopy and cyclic voltammetry.[11] The chemists showed, for example, that the complex [Re$_2$Cl$_8$]$^{2-}$ is stable in the basic ionic liquid and undergoes a one-electron reversible reduction process at a glassy-carbon electrode to form the elusive [Re$_2$Cl$_8$]$^{3-}$ ion.

9.4 ORGANOMETALLIC COMPOUNDS

Organometallic compounds are widely used as reactants and catalysts in organic synthesis. Because these compounds are polar, and many of them ionic, they are generally soluble in ionic liquids. As a consequence, there have been numerous studies of organometallic reactions and organometallic catalysis in ionic liquids over recent years (Chapters 8, 10 and 11).

Early studies of organometallic compounds in ionic liquids focused on catalysis, notably on the catalytic dimerizations of alkenes using nickel complexes (Section 8.8) and ethylene polymerization using an organo-titanium Ziegler–Natta catalyst (Section 10.22).

In 1996, Singer and colleagues showed for the first time that ionic liquids are suitable reaction media for the formation of new carbon–carbon bonds in organometallic compounds.[12] They demonstrated that the Friedel–Crafts acylation of ferrocene by acetic anhydride could be carried out in acidic [C$_2$mim]I-AlCl$_3$ (Scheme 9.2). The acidic chloroaluminate ionic liquid contains the Lewis acid [Al$_2$Cl$_7$]$^-$ which catalyses the reaction. The ionic liquid therefore acts both as a solvent and as a catalyst.

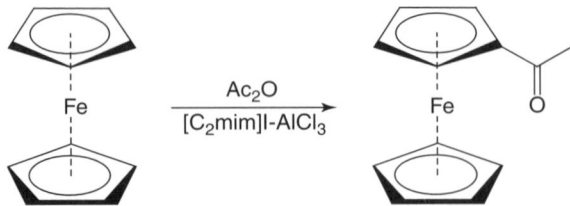

Scheme 9.2   Friedel–Crafts acylation of ferrocene.

Ferrocene, Fe(η-C$_5$H$_5$)$_2$, a compound consisting of two cyclopenta-dienyl rings that bind a central iron atom in a sandwich structure, is a classic example of an organometallic compound. Sandwich complexes with the general formula [Fe(C$_5$H$_5$)(arene)]$^+$ (**9.2**) can be generated from ferrocene in acidic [C$_4$mim]Cl-AlCl$_3$. The chloroaluminate ionic liquid can also be used to synthesize other transition metal-arene complexes such as the sandwich complex [Cr(arene)$_2$]$^+$ (**9.3**) and the half-sandwich complex [Mn(CO)$_3$(arene)]$^+$ (**9.4**).[13]

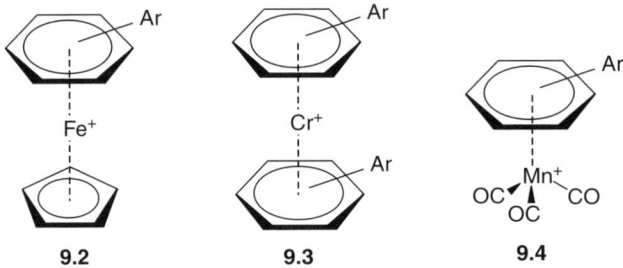

Basic chloroaluminate ionic liquids have also been used to prepare organometallic compounds. For example, Kozhevnikov and colleagues showed that the reaction of basic [C$_2$mim]Cl-AlCl$_3$ with a mixture of PtCl$_2$ and PtCl$_4$ and ethylene yielded a *N*-heterocyclic carbene complex of Pt(II).[14] During this reaction, the acidic proton in the 2-position of the imidazolium cation is removed under the basic conditions to yield 1-ethyl-3-methylimidazol-2-ylidene (**9.5**).

9.5

Carbenes are neutral two-coordinate carbon compounds with two non-bonding electrons. The >C: biradical moiety in carbenes can be stabilized by coordination with a metal to form a metal carbene complex. Transition metal complexes with *N*-heterocyclic carbene (NHC) ligands are known as persistent carbenes. These carbenes are stable and can be isolated, unlike normal carbenes, such as methylene ($H_2C:$), which are highly reactive and short lived. Persistent carbenes are therefore also known as stable carbenes.

Transition metal complexes with NHC ligands generated from ionic liquids are proving to be efficient catalysts for various reactions in organic chemistry, including the Heck and Suzuki coupling reactions (Chapter 11). The Heck and Suzuki reactions both employ organopalladium catalysts.

Cavell and co-workers have synthesized "unusually stable" NHC-metal-hydride complexes of nickel and palladium under mild conditions by insertion of the metal atom into the C2–H bond of cations in imidazolium ionic liquids like [C_1mim][BF_4] (Scheme 9.3).[15] The products are triscarbene-metal-hydrido complexes such as **9.6**. The authors suggested that these NHC-metal-hydride complexes may prove to be an efficient route for the generation of active catalytic species under mild conditions.

One of the technical problems of using organometallic catalysts dissolved in ionic liquids is recovery of the products from the ionic liquids using an organic extracting solvent without extracting some of the catalyst at the same time. If some of it is extracted, the catalyst becomes depleted each time the ionic liquid is recycled.

Such is the case with metal salen complexes dissolved in ionic liquids. Salen is a ligand prepared by the condensation of salicylic aldehyde and ethylene diamine. Metal-salen complexes catalyse reactions such as the epoxidation of alkenes with high activity and enantioselectivity not only in ionic liquids but also in conventional organic solvents (Section 8.9).

Baleizão and co-workers showed that metal salen complexes can be modified to make them insoluble in conventional organic solvents but miscible with ionic liquids.[16] The team covalently anchored a vanadyl salen complex to an imidazolium cation. The resulting complex (**9.7**) is

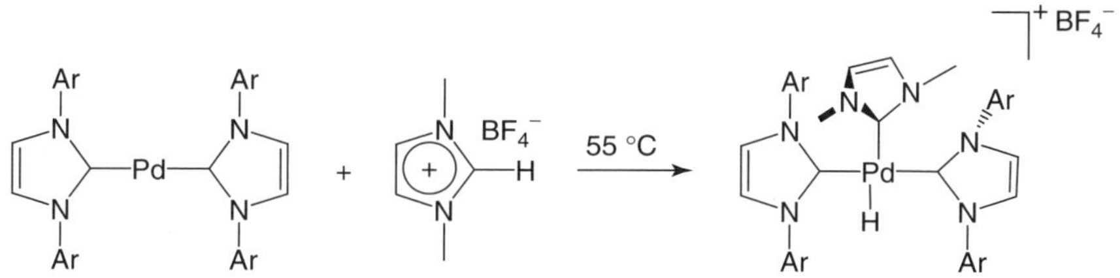

9.6

Scheme 9.3 Addition of an imidazolium salt to a palladium complex.

insoluble in hexane and diethyl ether but completely miscible with [C_4mim][PF_6]. The group demonstrated that the complex could be used as a highly active and reusable catalyst for the cyanosilylation of aldehydes in the ionic liquid. The chemists were able to reuse the ionic liquid–catalyst mixture up to six times without any detectable decrease in activity.

9.7

Organometallic ionic liquids with catalytically active anions are also known. In 2001, Dyson and co-workers reported the preparation of [C_4mim][$Co(CO)_4$] by the metathesis reaction of [C_4mim]Cl and Na[$Co(CO)_4$].[17] Transition metal carbonyl compounds are widely used in catalysis but only two are liquids: $Ni(CO)_4$ and $Fe(CO)_5$. The ionic liquid [C_4mim][$Co(CO)_4$] is a viscous liquid at temperatures down to 0 °C. Dyson's group showed that the ionic liquid catalyses the debromination of 2-bromoketones.

Dyson has also exploited the thermal stability of ionic liquids to synthesize transition metal carbonyl cluster compounds.[13] He carried out the thermolysis of $Ru_3(CO)_{12}$ in [C_4mim][BF_4] at 250 °C to yield a mixture of [$Ru_6H(CO)_{18}$]$^-$ and [$H_2Ru_{10}(CO)_{25}$]$^{2-}$. He also prepared [$Os_{10}C(CO)_{24}$]$^{2-}$ in high yield from $Os_3(CO)_{10}(NCCH_3)_2$.

In a separate development, Nockemann and co-workers prepared a hydrophobic task-specific ionic liquid that dissolves various metal oxides and metal hydroxides to form polynuclear metal complexes.[18] The task-specific ionic liquid, [Hbet][NTf_2], consists of betainium cations and [NTf_2]$^-$ anions. Betainium, [Hbet]$^+$, is the protonated form of betaine, $(CH_3)_3N^+CH_2COO^-$, a zwitterion that is also known as trimethylglycine.

The zwitterionic nature of the betaine ligand and the weakly coordinating ability of the [NTf_2]$^-$ anion enabled polynuclear metal betaine bistriflamide complexes with various crystal structures to be prepared. For example, the crystal structure of one of the compounds, [$Co_3(bet)_8(Hbet)_2(H_2O)_2$][$NTf_2$]$_{10}$[Hbet]$_2$, consists of discrete trimeric [$Co_3(bet)_8(Hbet)_2(H_2O)_2$]$^{8+}$ units, eight non-coordinating [NTf_2]$^-$ anions as counter ions, and two ion pairs of the ionic liquid [Hbet][NTf_2].

Nockemann's team prepared the compound by dissolving $Co(OH)_2$ in [Hbet][NTf_2] and water. After stirring under reflux, the water was removed by evaporation and the resulting purple crystals of the compound separated by filtration. The team also synthesized the compounds [Mn$_4$(bet)$_{10}$(H$_2$O)$_4$][NTf$_2$]$_8$ and [Ni$_5$(bet)$_{12}$(H$_2$O)$_6$][NTf$_2$]$_{10}$ in a similar way from MnO and NiO, respectively. The manganese(II) compound has discrete tetrameric cation units whereas the nickel(II) compound has a pentameric structure.

In 2002, Li and co-workers reported the first example of the use of an ionic liquid as a solvent to prepare a coordination polymer with an extended two-dimensional network structure.[19] The copper(I) network, [Cu(bpp)][BF$_4$], was prepared by the solvothermal reaction (Section 9.14) of Cu(NO$_3$)$_2 \cdot$3H$_2$O and 1,3-bis(4-pyridyl)propane (bpp) in [C$_4$mim][BF$_4$]. The group carried out the solvothermal reaction by sealing the reaction mixture and solvent in a glass tube and heating it at 140 °C for three days. The product appeared as column-like orange crystals. Single-crystal X-ray diffraction revealed that [Cu(bpp)][BF$_4$] consists of stacked parallel layers of [Cu(bpp)]$^+$ cations with [BF$_4$]$^-$ anions located between the layers.

Solvothermal reactions employ a solvent and are carried out in a closed vessel at elevated temperatures and, in general, high pressures. If the solvent is water, the process is known as "hydrothermal." When the solvent is an ionic liquid the technique is sometimes referred to as "ionothermal" (Section 9.14).

In 2006, Liao and co-workers described the use of the ionic liquid [C$_2$mim]Br for the ionothermal synthesis of a coordination polymer with a three-dimensional framework.[20] The framework of the polymer, [C$_2$mim][Cd(btc)], where btc is benzene-1,3,5-tricarboxylate, consists of [Cd(btc)]$^-$ anions with [C$_2$mim]$^+$ cations filling the void spaces.

All the examples of organometallic compounds described above contain cationic or anionic transition metal (d-block) complexes. Research has also been carried out on the use of ionic liquids to prepare organometallic compounds containing non-transition metals. For example, in 2006, Babai and Mudring described the use of ionic liquids to prepare three crystalline compounds containing anionic alkaline earth metal complexes.[21] The two chemists synthesized the complexes by heating mixtures of an alkaline earth metal iodide, CaI$_2$, SrI$_2$ or BaI$_2$ with an ionic liquid containing 1,1-*N*-methyl-*N*-propylpyrrolidinium (mppry) cations and [NTf$_2$]$^-$ anions.

The products, [mppyr]$_2$[Ca(NTf$_2$)$_4$], [mppyr]$_2$[Sr(NTf$_2$)$_4$] and [mppyr]$_2$[Ba(NTf$_2$)$_3$], were, according to the authors, the first examples of homoleptic alkaline earth metal bistriflamide complexes. In homoleptic complexes, all the ligands are identical. Alkaline earth metal bistriflamide compounds such as [Ca(H$_2$O)$_4$(NTf$_2$)$_2$], which have two different types of ligands, are known as heteroleptic compounds.

Mudring and co-workers also examined the crystal structures of rare earth (lanthanide) f-elements coordinated with $[NTf_2]^-$ ions. In 2005, they reported the structural characterization of the ytterbium complex $[Yb(NTf_2)_4]^{2-}$ obtained by the interaction of YbI_2 and the ionic liquid [mppyr][NTf_2].[22] The complex crystallizes with four formula units in the unit cell. The researchers concluded that the $[NTf_2]^-$ anion is not as non-coordinating as often believed. In the absence of a stronger coordinating ligand, $[NTf_2]^-$ coordinates to a metal cation in solution.

Following on from this work, Mudring showed that ionic liquids can be used to engineer the crystal structure of organometallic compounds containing rare earth complex anions.[23] They prepared single crystals of $[mppyr][NdI_6]$ by the reaction of neodymium iodide, NiI_3, with [mppyr][NTf_2]. The compound $[mppyr][NdI_6]$ crystallizes with eight formula units in its unit cell.

On replacing the propyl side chain of the $[mppyr]^+$ cation with butyl, the researchers obtained a product that crystallizes with four formula units in its unit cell. The ionic liquid in this case contains 1-butyl-1-methylpyrrolidinium, $[bmpyr]^+$, cations and $[NTf_2]^-$ anions. The compound, $[bmpyr]_4[NdI_6][NTf_2]$, which is produced by the reaction of the ionic liquid with NiI_3, incorporates the cation and the anion of the ionic liquid in its crystal structure. The ionic liquid cation with the butyl side chain is larger than that with the propyl side. Optimal packing of the larger cation in the solid requires incorporation of one formula unit of the ionic liquid solvent in its crystal structure. The formula might therefore be rewritten as $[bmpyr]_3[NdI_6].[bmpyr][NTf_2]$.

The work illustrates how a small change in an ionic liquid solvent can have an impact on crystal structure. It also points to the potential use of ionic liquids for crystal engineering. Rogers and co-workers observed, for example, that ionic liquids are complex crystallization solvents that are generating a "large number of variable, complex, and captivating results", particularly for organometallic coordination compounds.[24] They suggested, in their article on the topic, that the growing under-standing of the interactions of ionic liquids with, for example, metals and ligands promises "intimate control over crystallization processes."

9.5 NUCLEAR CHEMISTRY

The potential application of ionic liquids for nuclear fuel reprocessing and nuclear waste treatment has sparked several investigations into the chemistry of radioactive actinide elements, notably uranium, in ionic liquids.

The study and behaviour of the uranium in ionic liquids dates back to 1982 when De Waele and colleagues showed that uranium(IV) chloride is soluble in acidic [C$_4$py]Cl-AlCl$_3$.[25] They showed that uranium(IV) could be electrochemically reduced or oxidized in the ionic liquid to give uranium(III) or uranium(V), respectively.

Uranium(IV) oxide, UO$_2$, containing the fissile ^{235}U isotope is a major component of spent nuclear fuel. Small amounts of ^{239}Pu and other fission products are also present in the spent fuel. Commercial reprocessing of the spent fuel uses the Purex process, during which the irradiated fuel is dissolved in hot concentrated nitric acid to convert the uranium, plutonium and other fission products into nitrates. The uranyl and plutonium nitrates are extracted and separated by solvent extraction. They are then converted into oxides for reuse as nuclear fuels.

In the 1990s, chemists at British Nuclear Fuels in Sellafield, UK, working in collaboration with chemists at Queen's University of Belfast, began investigating the possibility of using ionic liquids such as [C$_4$mim][NO$_3$] to separate the components of the spent fuel in a manner that was not only more compact and cost-effective than the commercial process but would also minimize the waste produced. However, UO$_2$ and plutonium(IV) oxide, PuO$_2$, are insoluble in pure nitrate ionic liquids, even at elevated temperatures, without the addition of an oxidant such as concentrated nitric acid. When an oxidant is added, the U(IV) and Pu(IV) species are converted into complexes containing U(VI) and Pu(VI), which are soluble in nitrate ionic liquids.

The Purex process uses tri-*n*-butylphosphate (TBP) diluted with kerosene as an extractant to remove the uranyl and plutonium nitrates from the nitrate solution. The extractant CMPO [octyl(phenyl)(*N,N*-diisobutylcarbomylmethyl)phosphine oxide] (**9.8**) is also used in combination with TBP to partition actinides with different oxidation states from nitric acid solutions into organic solvents.

9.8

Rogers and co-workers examined the partitioning of Am(III), Th(IV), U(VI) and Pu(VI) species between aqueous systems and hydrophobic ionic liquids, such as [C$_4$mim][PF$_6$], using CMPO and mixtures of

CMPO and TBP as extractants.[26] They concluded that the coordination and extraction of actinides into ionic liquid solvents may be more efficient than into conventional solvents.

In 2002, Seddon and co-workers reported an investigation of the radiochemical stability of ionic liquids.[27] The group subjected the imidazolium ionic liquids [C$_2$mim]Cl, [C$_6$mim]Cl and [C$_4$mim][NO$_3$] to doses of radiation. The results suggested that the radiation stability of imidazolium ionic liquids is similar to that of benzene and better than that of mixtures of TBP and kerosene under similar irradiation conditions.

The chemistry of trivalent lanthanide ions is similar to that of trivalent actinide ions in aqueous solutions. Lanthanide(III) ions have therefore been considered as stand-in ions for studying the chemistry of radioactive actinide(III) ions in ionic liquids.[28] However, in 2007, Stumpf and co-workers showed that the chemical behaviour of the two types of ions is "dramatically" different in ionic liquids.[29] The team used time-resolved laser fluorescence spectroscopy to study the coordination structure and chemistry of the lanthanide europium(III) and the actinide curium(III) in [C$_4$mim][NTf$_2$].

The spectroscopic data indicated comparable coordination for the Eu(III) and Cm(III) cations. However, addition of copper(II) ions to a solution of a Eu(III) salt in the ionic liquid quenched the fluorescence emission of Eu(III) whereas Cu(II) ions did not quench the fluorescence emission of Cm(III) ions dissolved in the ionic liquid. The different chemistry could prove useful in separating lanthanides and actinides when nuclear fuel is reprocessed, according to the team.

9.6 NANOPARTICLES

Particles with at least one dimension on the nanometre (10^{-9} m) scale are known as nanoparticles. The ability to control the chemical and physical characteristics of inorganic nanoparticles of materials such as carbon, metals and semiconductors can lead to improved electrical conductivity, catalytic, magnetic and optical activity and other properties of the bulk materials composed of these nanoparticles. These properties depend to a large extent on the size, shape, and type of the nanoparticles and on the use of materials, where necessary, to stabilize them and prevent aggregation.

Nanostructured materials with well-defined particle shapes and sizes are now being exploited or developed to enhance the performance of gas sensors and other industrial monitoring devices and to produce better

fuel-cells and lighter weight and longer lasting batteries. The high sur-
face-to-volume ratio of transition metal nanoparticles makes them
particularly attractive for use in catalysis (Section 8.6).

Inevitably, many chemists have realised over the past few years that
ionic liquids with their unique chemical and physical properties open up
opportunities for the synthesis of materials with novel nanostructures
and for the development of new and more efficient routes for the
manufacture of nanomaterials. The following five sections outline some
of the research on this topic over the past few years.

9.7 CARBON NANOTUBES

In one of the earliest papers linking ionic liquids to nanoparticles,
published in 2003, Fukushima, Aida and co-workers described how
ionic liquids and carbon nanotubes can be combined to form non-
volatile thermally stable gels that might potentially be used to manufacture coating materials, antistatic materials and novel electronic
devices.[30]

The group prepared their "bucky gels" by grinding up suspensions of
high-purity single-walled carbon nanotubes in ionic liquids with various
imidazolium cations. Transmission electron microscopy revealed that
the gels consist of untangled bundles of the nanotubes that physically
cross-link due to the cation–π interactions between the nanotube surfaces and the imidazolium ions.

The researchers also used a bucky gel formed with a polymerizable
ionic liquid, 1-(4-acryloyloxybutyl)-3-methylimidazolium hexafluorophosphate (**9.9**), to prepare a highly electroconductive plastic material
reinforced with single-walled carbon nanotubes. Addition of just 4% by
weight of the nanotubes enhanced the mechanical properties of the
polymer, such as hardness, by around 400% owing to the strong inter-
facial interaction between the nanotubes and the ionic liquid polymer.

9.9

In 2006, Liu and co-workers showed that the solubility of multiwalled carbon nanotubes modified with an imidazolium ionic liquid
could be reversibly switched between aqueous and organic solvents by
exchanging the anions of the ionic liquid using electrolyte solutions.[31]

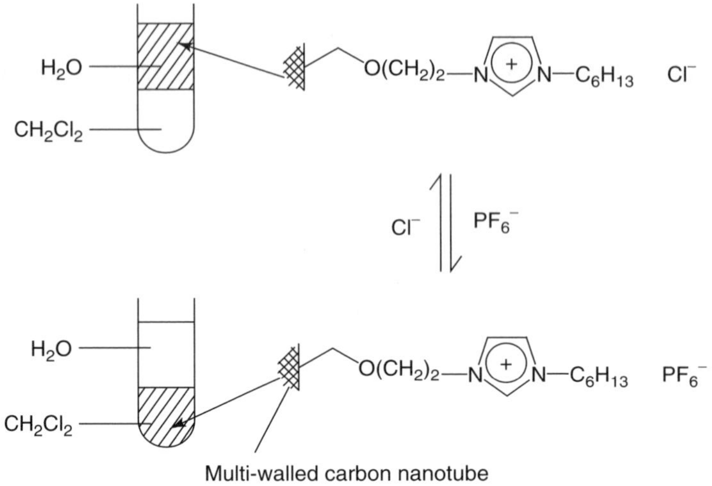

Figure 9.1   Carbon nanotubes switch between phases.

For example, nanotubes coupled to an ionic liquid with the hydrophobic $[PF_6]^-$ anion formed a stable dispersion in the lower CH_2Cl_2 phase of a biphasic CH_2Cl_2/H_2O system (Figure 9.1). On addition of an aqueous solution containing a salt such as NaCl, the ionic liquid modified nanotubes switched to the upper aqueous phase. The authors suggested that the work could create new opportunities for practical applications of carbon nanotubes in sensors and actuators and in biological and biomedical science.

9.8   METAL AND SEMICONDUCTOR NANOPARTICLES

Ionic liquids have been exploited for the preparation and stabilization of transition metal nanoparticles for use as catalysts in organic chemistry. For example, Dupont and co-workers prepared iridium and rhodium nanoparticles using an ionic liquid.[32] They used molecular hydrogen for the reduction of iridium(I) and rhodium(III) compounds in $[C_4mim][PF_6]$ to form nanoparticles of the two metals with diameters in the range 2–3 nm. The team showed that the nanoparticles are efficient catalysts for the hydrogenation of arenes.

More recently, Dupont and colleagues showed that the high thermal stability of ionic liquids can be exploited for the preparation of transition metal nanoparticles by *in situ* thermal decomposition.[33] The group dispersed $[Co_2(CO)_8]$ in $[C_{10}mim][NTf_2]$ and other imidazolium ionic liquids and heated the mixture at 150 °C for over one hour.

Transmission electron micrography revealed that the resulting black solution contained "naked" ligand-free cobalt nanoparticles with cubic or other shapes depending on the type of ionic liquid used and the reaction time. The team concluded that, in principle, the shapes of the nanoparticles can be tuned by varying the length of the imidazolium side chains, the type of ionic liquid anion and the reaction conditions.

Metal nanoparticles often aggregate when dispersed in a solvent. Nanoparticles are therefore frequently stabilized against aggregation by protecting their surfaces with what is known as a capping agent.

In 2005, Tatumi and Fujihara showed that a zwitterionic liquid, in which the zwitterion consists of an imidazolium cation linked to a sulfonate anion, could be used as a capping agent to stabilize gold nanoparticles.[34] The two scientists showed that the "zwitter-gold" nanoparticles **9.10** are soluble and stable in the ionic liquid $[C_2mim][BF_4]$ as well as in aqueous electrolyte solutions. They suggested that the use of ionic liquid- and water-soluble nanoparticles opens up "interesting perspectives" in catalysis and electronic and bioanalytical applications.

9.10

A year later, Wang and Yang reported that oleic acid, $C_{17}H_{33}COOH$, is an effective capping agent for the synthesis of silver and platinum nanoparticles in imidazolium-based ionic liquids.[35] The two researchers prepared noble metal nanoparticles by heating a noble metal precursor with the acid in $[C_4mim][NTf_2]$. They showed that the use of the capping agent not only stabilizes the silver and platinum nanoparticles but also causes the nanoparticles to settle in the ionic liquid. The ionic liquid can then be readily removed, leaving the settled nanoparticles to be dispersed in a conventional organic solvent such as hexane.

Metal nanoparticles required for catalysis are often prepared in the aqueous phase and then transferred to an organic phase. Their solubility in organic phases can sometimes be enhanced by capping the nanoparticle surfaces with a thiol or amine. However, capping limits the catalytic activity of the nanoparticles.

In 2004, Wei and colleagues demonstrated that ionic liquids can be used as the organic phase for the aqueous-to-organic phase transfer of well-defined uncapped metal nanoparticles.[36] The team transferred gold nanoparticles and gold nanorods from an aqueous solution to the water immiscible ionic liquid $[C_4mim][PF_6]$ by vigorously shaking a biphasic mixture of the two liquids. Before-and-after transmission electron microscopy images revealed that the size of the nanoparticles did not change during phase transfer. Similarly, images of gold nanorods showed that the shape of the nanorods is preserved following transfer to the ionic liquid phase.

Ionic liquids have also been exploited to synthesize nanoparticles of semiconducting elements. Zhu and co-workers used microwave-assisted synthesis to prepare tellurium nanorods and nanowires in $[C_4py][BF_4]$.[37] Tellurium is a p-type semiconductor that exhibits useful properties such as photoconductivity and high piezoelectricity. It has a multitude of uses in the semiconductor, electronics and other industries. Zhu's group prepared the Te nanoparticles by reducing TeO_2 with $NaBH_4$ in the ionic liquid at 180 °C in a microwave oven. By controlling the experimental parameters, it was possible to synthesize either nanorods or nanowires.

9.9 ALLOY NANOPARTICLES

Nanoparticles of alloys exhibit different catalytic, magnetic, optical and other chemical and physical properties to those of single-metal nanoparticles. They are potentially useful for applications such as chemical and biochemical sensing, catalysis, optoelectronic devices, medical diagnostics and therapeutics.

The properties of alloy nanoparticles depend not only on the size and shape of the nanoparticles but also on their composition. Wang and Yang exploited the high thermal stability of ionic liquids to synthesize cobalt platinum nanorods with different CoPt compositions.[38] They prepared the bimetallic nanorods by heating Pt and Co precursors and the capping reagent cetyltrimethylammonium bromide at 350 °C in $[C_4mim][NTf_2]$.

The use of a stabililizing agent to prepare single-metal and alloy nanoparticles in ionic liquids can be avoided by using a technique known as sputter deposition. The technique ejects atoms from a solid surface of a target by the impact of energetic gaseous ions, typically Ar^+ ions.

In 2008, Torimoto and co-workers used the technique to synthesize gold-silver alloy nanoparticles in an ionic liquid.[39] The group injected the bimetallic nanoparticles into $[C_4mim][PF_6]$ by sputtering pure gold and silver simultaneously from a foil target that contained separate surface domains for each metal. The process caused the two metals to coalesce into bimetallic nanoparticles in the gas phase before being deposited into the ionic liquid. By varying the area ratio of the pure gold and silver domains in the sputtering foil targets, the researchers were able to control the chemical composition and optical properties of the alloy nanoparticles.

9.10 METAL OXIDE NANOPARTICLES

Ionic liquids have also been used to synthesize various metal oxide nanoparticles. For example, Zhou and Antonietti have prepared crystalline titanium dioxide, TiO_2, nanoparticles by the hydrolysis of titanium tetrachloride, $TiCl_4$, in [C$_4$mim][BF$_4$] at $80\,°C$.[40] The nanocrystals self-assembled in the ionic liquid to form a new sponge-like TiO_2 structure with a high surface area and a narrow distribution of pore sizes. The authors suggested that such materials are potentially useful for applications such as solar energy conversion, catalysis and optoelectronic devices.

Farag and Endres prepared alumina, Al_2O_3, nanoparticles by adding aluminium chloride, $AlCl_3$, to pyrrolidiniuim- and imidazolium-based ionic liquids containing [NTf$_2$]$^-$ anions.[41] Addition of dry $AlCl_3$ to the ionic liquids resulted in biphasic mixtures that became monophasic at around $80\,°C$. Addition of water to the upper phase of the biphasic mixtures or to the homogeneous solutions yielded the hydrolysis product $AlCl_3 \cdot 6H_2O$. After separation from the ionic liquid, the product was heated to various temperatures to give different forms of alumina, including porous alumina with average pore sizes of around 10 nm.

Zheng and colleagues explored the effect of ionic liquid cations on the shape of zinc oxide nanoparticles.[42] They investigated the formation of ZnO nanoparticles, nanorods and nanowires using three imidazolium-based ionic liquids containing [BF$_4$]$^-$ anions. They concluded that the structure of the ionic liquid cations plays a crucial role in determining the morphology of the final product. For example, hydrogen bonding between the hydrogen atom at position 2 of the imidazole ring and the ZnO oxygen atoms promotes the directional growth of one-dimensional ZnO nanostructures. Longer alkyl chains at position 1 hinder the ZnO nanostructures from growing longer.

9.11 OTHER NANOPARTICLES

Ionic liquids have been used to prepare nanoparticles not only of pure metals, alloys and metal oxides but also of other types of metallic compounds.

Bühler and Feldmann, for instance, exploited the chemical and thermal stability of ionic liquids for the microwave-assisted synthesis of nanocrystals of a rare earth doped metal phosphate.[43] These nanocrystalline-doped phosphates are attracting a lot of interest as luminescent materials for optical systems, displays and fluorescent labelling for biomedical applications. However, to produce high quality luminescent nanocrystals, toxic materials are often used. Furthermore, the syntheses typically require temperatures in the range 150–250 °C to minimize lattice defects.

The two chemists showed that the ionic liquid $[N_{1\ 4\ 4\ 4}][NTf_2]$ can be used as a co-solvent with an organic solvent for the microwave-assisted synthesis of luminescent nanocrystals of lanthanum phosphate doped with cerium and terbium: $LaPO_4$:Ce,Tb. The reaction was carried out at 70 °C using a phosphate precursor such as phosphoric acid, H_3PO_4, and the hydrated rare earth chlorides $LaCl_3 \cdot 6H_2O$, $CeCl_3 \cdot 7H_2O$ and $TbCl_3 \cdot 6H_2O$. The authors noted that the absence of harmful solids, surface modifiers and solvents in the nanocrystals "could be particularly beneficial for biomedical applications."

In a separate development, Larionova, Guari, Sangregorio and co-workers used various imidazolium-based ionic liquids as stabilizing agents and solvents for the preparation of colloidal solutions containing nanoparticles of cyano-bridged coordination polymers.[44] These polymers have potentially useful magnetic, optical, photo-switchable and intercalation properties.

The polymer nanoparticles synthesized by the team have the general formula $Cu_3[Fe(CN)_6]_2/[C_n mim][An]$ (where $n = 2$, 4; $An = BF_4$, Cl). The colloidal solutions consisted of these stabilized nanoparticles dispersed in the ionic liquid solvent $[C_n mim][An]$. The solutions, which proved "exceptionally stable" according to the authors, were prepared by mixing solutions of $[C_4 mim]_3[Fe(CN)_6]$ and $[Cu(H_2O)_x][BF_4]_2$ in the ionic liquid. The size of the nanoparticles was varied by changing reaction conditions such as temperature.

9.12 MICROSPHERES

Microspheres are spheres with diameters ranging from a few to thousands of micrometres. They are used for a wide range of applications. For example, metal alloy microspheres are used for metallurgical and electronic applications. Metal oxide microspheres are used as catalyst carriers, in high-tech ceramics and as abrasion resistant materials for grinding other materials.

Ionic liquids have been shown to be useful media, stabilizers and morphology templates for the synthesis of metal oxide microspheres. In one piece of work in this field, Nakashima and Kimizuka synthesized hollow titania, TiO_2, gel microspheres in an ionic liquid using a sol–gel process.[45]

The sol–gel process typically converts a colloidal solution, *i.e.* a sol, into a gel. Gels have highly porous three-dimensional network structures and properties intermediate between a solid and a liquid.

Nakashima and Kimizuka prepared the TiO_2 gel by vigorously stirring a solution of titanium tetrabutoxide, $Ti(OC_4H_9)_4$, in an organic solvent with $[C_4mim][PF_6]$ containing a trace of water. The resulting hollow TiO_2 gel microspheres were removed by filtration and dried. The two researchers reported that gold nanoparticles or dye molecules could be incorporated into the hollow microspheres during the process by adding the nanoparticles or molecules to the organic solution of $Ti(OC_4H_9)_4$. The authors suggested that the single-step process provides a simple and general route for the design and synthesis of a new family of hollow inorganic capsules with functional organic or inorganic components that might prove useful for photocatalysis.

Microspheres of tin dioxide, SnO_2, have also been synthesized in an ionic liquid. The oxide has many uses, *e.g.* in the manufacture of glasses and ceramic glazes and, because it is an n-type semiconductor, it is also used in various electronic applications.

Dong, Liu and co-workers tested various imidazolium and pyridinium ionic liquids for the microwave-assisted synthesis of SnO_2.[46] They prepared, for example, SnO_2 microspheres with an average diameter of 2.5 μm by the hydrothermal reaction of $SnCl_4 \cdot 5H_2O$ in an aqueous solution of sodium hydroxide and $[C_4mim][BF_4]$. The authors suggested that the $[BF_4]^-$ anions interact electrostatically with the SnO_2 crystallites and, therefore, act as templates for the growth of the microspheres.

The team carried out the hydrothermal reaction by heating the aqueous solution in a microwave oven at 160 °C. After cooling, an aqueous suspension of the microspheres was collected by centrifugation. The suspension was then washed with water and the microspheres obtained by drying in the suspension in vacuum.

In related work, Chen and co-workers used "ionic liquid-assisted hydrothermal synthesis" to prepare molybdenum disulfide, MoS_2, microspheres with uniform sizes and mean diameters of about 2.1 μm.[47] The team prepared the microspheres by adding the ionic liquid $[C_4mim][BF_4]$ to an aqueous solution of sodium molybdate and thioacetamide, CH_3CSNH_2. The mixture was then heated in a tightly sealed autoclave. The ionic liquid acted as a template for the microspheres. The hydrophobic butyl chains of the $[C_4mim]^+$ cations caused the formation of vesicles with molybdate ions, MoO_4^{2-}, on their surface. The cations and MoO_4^{2-} combined by electrostatic interactions. The vesicles provided nucleation domains for the hydrothermal reaction between MoO_4^{2-} and H_2S, which was generated *in situ* by the thermal decomposition of CH_3CSNH_2.

9.13   SOL–GEL SYNTHESIS OF POROUS MATERIALS

The potential of using ionic liquids to prepare porous materials was demonstrated as early as 2000 when Dai and co-workers synthesized silica, SiO_2, aerogels using $[C_2mim][NTf_2]$ as a solvent.[48]

Aerogels consist of a gas dispersed in a continuous solid three-dimensional network. On drying, the pores shrink to form a xerogel. The low density, thermal conductivity and insulating properties of aerogels combined with their high surface areas make these materials potentially attractive for use in separation technology, catalysis, thermal insulation, optical devices, sensors and other applications.

The conventional method of preparing silica aerogels employs sol–gel synthesis. A tetraalkylorthosilicate such as $(CH_3O)_4Si$ is hydrolysed in an alcoholic aqueous solution. The colloidal silica particles form by condensation. The solvent is then allowed to evaporate, resulting in shrinkage of the gel and formation of a stable sol–gel network. If this aging process is too short, the gel network can become unstable and collapse. If the process is too long, the pores in the gel shrink to the nanoscale pore dimensions of a xerogel. Control of solvent evaporation and aging time are therefore critical in the synthesis of SiO_2 aerogels. Shrinkage can also be prevented by using supercritical CO_2 to remove the solvent.

Dai's group recognized that ionic liquids do not evaporate because of their negligible vapour pressures and might therefore be suitable solvents for preparing SiO_2 aerogels. The researchers showed that a stable aerogel can be formed by the hydrolysis of $(CH_3O)_4Si$ in $[C_2mim][NTf_2]$ using formic acid as a catalyst. After aging the aerogel network for three weeks at room temperature, the ionic liquid entrapped in its pores was extracted by refluxing the material in acetonitrile. The extraction process did not result in collapse or any visible shrinkage of the SiO_2 network.

9.14   IONOTHERMAL SYNTHESIS OF POROUS MATERIALS

Much of the attention on the use of ionic liquids to prepare porous materials has focused on the ionothermal synthesis of zeolites and their synthetic analogues. Naturally occurring and synthetic zeolites are aluminosilicate materials with porous framework structures that can accommodate water and other molecules and also cations. The materials, which are also known as molecular sieves, are used as hosts for catalysts in industrial processes, water purification and softening, laundry detergents, soil treatment and the preparation of medical-grade oxygen. Zeolite analogues include aluminophosphates and other synthetic framework materials.

Synthetic microporous materials such as aluminophosphates are typically prepared by solvothermal synthesis, notably hydrothermal synthesis, which employs water as a solvent. The reaction mixtures for the syntheses consist of the framework precursors, *e.g.* aluminium and phosphate compounds, an organic structure-directing agent (also known as a template), a mineralizer to promote crystallization and the solvent. Conventional heating or microwave heating is used to heat the mixture in a sealed vessel under "autogenous" pressure, *i.e.* pressure generated by evaporation of the solvent inside the vessel. Compared with conventional heating, microwave heating leads to rapid crystal growth rate and higher structural selectivity of the product.

Ionothermal syntheses, *i.e.* solvothermal syntheses that employ an ionic liquid as a solvent, do not result in autogenous pressure because the solvents have negligible vapour pressures. The syntheses can therefore be carried out at ambient pressures. In addition, the ionic liquid not only acts as a solvent for the synthesis but its cations also provide a template around which the inorganic frameworks are ordered. The ionic liquid therefore removes competition between template–framework and solvent–framework interactions that are known to occur in hydrothermal syntheses.

One of the advantages of using ionic liquids for ionothermal syntheses is their thermal stability. The ionic liquid $[C_2mim]Br$ has been used extensively for ionothermal syntheses because it is a relatively polar solvent that dissolves framework precursors. However, it is not as thermally stable as some other ionic liquids. It has been shown to decompose under vacuum at temperatures in excess of 220–260 °C. Ionothermal syntheses using this ionic liquid are therefore typically carried out at around 150 °C.

Ionic liquids with the $[NTf_2]^-$ anion are stable at temperatures over 500 °C but are hydrophobic. They are, consequently, poor solvents for inorganic compounds. Fluoride ions are often added in ionothermal synthesis to enhance the solubility of the inorganic starting materials in the ionic liquid. The ions also act as a "mineralizer" that promotes crystallization of the framework compounds.

In 2004, Morris and colleagues reported the ionothermal synthesis of aluminophosphate zeolite analogues using the ionic liquid [C$_2$mim]Br or a eutectic mixture of choline chloride and urea.[49] The group selected [C$_2$mim]Br as an ionic liquid because it dissolves the framework precursors, Al[OCH(CH$_3$)$_2$]$_3$ and H$_3$PO$_4$. The ionic liquid acted not only as a solvent but also as a template. The syntheses in the ionic liquid and eutectic mixtures were carried out at 150 and 180 °C, respectively.

Two years later, Xu and colleagues reported the microwave-assisted ionothermal synthesis of aluminophosphate and silicoaluminophosphate (SAPO) molecular sieves in [C$_2$mim]Br and [C$_4$mim]Br at 100–150 °C.[50] Once again, the ionic liquids acted as both solvents and templates for the syntheses. An aqueous solution of hydrofluoric acid, HF, was added as the mineralizer.

In a subsequent development, Yan and co-workers used microwave-assisted ionothermal synthesis and [C$_2$mim]Br as a solvent and template to coat copper-containing aluminium alloys with an aluminophosphate zeolite and also with a SAPO zeolite.[51] The SAPO coating exhibited "very good" anticorrosion properties whereas the aluminophosphate provided relatively little corrosion resistance to the alloy surface, according to the team.

Transition-metal-functionalized frameworks with potential applications in catalysis and gas storage can also be prepared by ionothermal synthesis. For example, Morris and co-workers have prepared three cobalt aluminophosphate zeolite frameworks by ionothermal synthesis using [C$_2$mim]Br as the solvent and template.[52] They prepared the materials by heating Al[OCH(CH$_3$)$_2$]$_3$, H$_3$PO$_4$, Co(OH)$_2$ and the mineralizer HF in an autoclave at 170 °C for three days.

In 2004, Kim and colleagues reported the first ionothermal synthesis of a three-dimensional metal-organic framework.[53] The framework was prepared by heating a mixture of 2,4,6-tris(4-pyridyl)-1,3,5-triazine (tpp) and Cu(NO$_3$)$_2$·3H$_2$O in [C$_4$mim][BF$_4$] in a sealed glass tube for two days. The deep violet crystals of [Cu$_3$(tpt)$_4$](BF$_4$)$_3$·(tpt)$_{2/3}$·5H$_2$O that resulted were characterized by single-crystal and X-ray powder diffraction, infrared spectroscopy, thermogravimetric analysis and elemental analysis. The structure proved to have large channels of around 5 Å in diameter filled with non-coordinating free tpt molecules, water and BF$_4^-$ anions.

REFERENCES

1. R. P. Swatloski, J. D. Holbrey and R. D. Rogers, *Green Chem.*, 2003, **5**, 361.
2. Z. Zhang, Y. Xie, W. Li, S. Hu, J. Song, T. Jiang and B. Han, *Angew. Chem. Int. Ed.*, 2008, **47**, 1127.
3. Z.-W. Wang and L.-S. Wang, *Appl. Catal., A*, 2004, **262**, 101.
4. D.-C. Chen and L.-S. Wang, *J. Chem. Eng. Data*, 2005, **50**, 616.
5. E. Amigues, C. Hardacre, G. Keane, M. Migaud and M. O'Neill, *Chem. Commun.*, 2006, 72.
6. E. J. Amigues, C. Hardacre, G. Keane and M. E. Migaud, *Green. Chem.*, 2008, **10**, 660.
7. R. J. Gale, B. Gilbert and R. A. Osteryoung, *Inorg. Chem.*, 1979, **18**, 2723.
8. C. L. Hussey and T. M. Laher, *Inorg. Chem.*, 1981, **20**, 4201.
9. I.-W. Sun, E. H. Ward and C. L. Hussey, *Inorg. Chem.*, 1987, **26**, 4309.
10. C. L. Hussey, *Pure & Appl. Chem.*, 1988, **60**, 1763.
11. S. K. D. Strubinger, I.-W. Sun, W. E. Cleland Jr. and C. L. Hussey, *Inorg. Chem.*, 1990, **29**, 993.
12. J. K. D. Surette, L. Green and R. D. Singer, *Chem. Commun.*, 1996, 2753.
13. P. J. Dyson, *Transition Met. Chem.*, 2002, **27**, 353.
14. M. Hasan, I. V. Kozhevnikov, M. R. H. Siddiqui, A. Steiner and N. Winterton, *J. Chem. Res. (S)*, 2000, 392.
15. N. D. Clement, K. J. Cavell, C. Jones and C. J. Elsevier, *Angew. Chem. Int. Ed.*, 2004, **43**, 1277.
16. C. Baleizão, B. Gigante, H. Garcia and A. Corma, *Tetrahedron. Lett.*, 2003, **44**, 6813.
17. R. J. C. Brown, P. J. Dyson, D. J. Ellis and T. Welton, *Chem. Commun.*, 2001, 1862.
18. P. Nockemann, B. Thijs, K. Van Hecke, L. Van Meervelt and K. Binnemans, *Cryst. Growth. Des.*, 2008, **8**, 1353.
19. K. Jin, X. Huang, L. Pang, J. Li, A. Appel and S. Wherland, *Chem. Commun.*, 2002, 2872.
20. J.-H. Liao, P.-C. Wu and W.-C. Huang, *Cryst. Growth Des.*, 2006, **6**, 1062.
21. A. Babai and A.-V. Mudring, *Inorg. Chem.*, 2006, **45**, 3249.
22. A.-V. Mudring, A. Babai, S. Arenz and R. Giernoth, *Angew. Chem. Int. Ed.*, 2005, **44**, 5485.
23. A. Babai and A.-V. Mudring, *Inorg. Chem.*, 2006, **45**, 4874.

24. W. M. Reichert, J. D. Holbrey, K. B. Vigour, T. D. Morgan, G. A. Broker and R. D. Rogers, *Chem. Commun.*, 2006, 4767.
25. R. De Waele, L. Heerman and W. D'Olieslager, *J. Electroanal. Chem.*, 1982, **142**, 137.
26. K. E. Gutowski, N. J. Bridges, V. A. Cocalia, S. K. Spear, A. E. Visser, J. D. Holbrey, J. H. Davis Jr and R. D. Rogers, in *Ionic Liquids IIIB: Fundamentals, Progress, Challenges, and Opportunities: Transformations and Processes*, ed. R. D. Rogers and K. R. Seddon, ACS Symposium Series 902, American Chemical Society, Washington DC, 2005, p. 33.
27. D. Allen, G. Baston, A. E. Bradley, T. Gorman, A. Haile, I. Hamblett, J. E. Hatter, M. J. F. Healey, B. Hodgson, R. Lewin, K. V. Lovell, B. Newton, W. R. Pitner, D. W. Rooney, D. Sanders, K. R. Seddon, H. E. Sims and R. C. Thied, *Green Chem.*, 2002, **4**, 152.
28. V. A. Cocalia, K. E. Gutowski and R. D. Rogers, *Coord. Chem. Rev.*, 2006, **250**, 755.
29. S. Stumpf, I. Billard, P. J. Panak and S. Mekki, *Dalton Trans.*, 2007, 240.
30. T. Fukushima, A. Kosaka, Y. Ishimura, T. Yamamoto, T. Takigawa, N. Ishii and T. Aida, *Science*, 2003, **300**, 2072.
31. B. Yu, F. Zhou, G. Liu, Y. Liang, W. T. S. Huck and W. Liu, *Chem. Commun.*, 2006, 2356.
32. G. S. Fonseca, A. P. Umpierre, P. F. P. Fichtner, S. R. Teixeira and J. Dupont, *Chem. Eur. J.*, 2003, **9**, 3263.
33. M. Scariot, D. O. Silva, J. D. Scholten, G. Machado, S. R. Teixeira, M. A. Novak, G. Ebeling and J. Dupont, *Angew. Chem. Int. Ed.*, 2008, **47**, 9215.
34. R. Tatumi and H. Fujihara, *Chem. Commun.*, 2005, 83.
35. Y. Wang and H. Yang, *Chem. Commun.*, 2006, 2545.
36. G.-T. Wei, Z. Yang, C.-Y. Lee, H.-Y. Yang and C. R. C. Wang, *J. Am. Chem. Soc.*, 2004, **126**, 5036.
37. Y.-J. Zhu, W.-W. Wang, R.-J. Qi and X.-L. Hu, *Angew. Chem. Int. Ed.*, 2004, **43**, 1410.
38. Y. Wang and H. Yang, *J. Am. Chem. Soc.*, 2005, **127**, 5316.
39. K. Okazaki, T. Kiyama, K. Hirahara, N. Tanaka, S. Kuwabata and T. Torimoto, *Chem. Commun.*, 2008, 691.
40. Y. Zhou and M. Antonietti, *J. Am. Chem. Soc.*, 2003, **125**, 14960.
41. H. K. Farag and F. Endres, *J. Mater. Chem.*, 2008, **18**, 442.
42. L. Wang, L. Chang, B. Zhao, Z. Yuan, G. Shao and W. Zheng, *Inorg. Chem.*, 2008, **47**, 1443.
43. G. Bühler and C. Feldmann, *Angew. Chem. Int. Ed.*, 2006, **45**, 4864.

44. J. Larionova, Y. Guari, A. Tokarev, E. Chelebaeva, C. Luna, C. Sangregorio, A. Caneschi and C. Guérin, *Inorg. Chim. Acta*, 2008, **361**, 3988.
45. T. Nakashima and N. Kimizuka, *J. Am. Chem. Soc.*2003, **125**, 6386.
46. W.-S. Dong, M.-Y. Li, C. Liu, F. Lin and Z. Liu, *J. Colloid Interfac. Sci.*, 2008, **319**, 115.
47. L. Ma, W.-X. Chen, H. Li, Y.-F. Zheng and Z.-D. Xu, *Mater. Lett.*, 2008, **62**, 797.
48. S. Dai, Y. H. Ju, H. J. Gao, J. S. Lin, S. J. Pennycook and C. E. Barnes, *Chem. Commun.*, 2000, 243.
49. E. R. Cooper, C. D. Andrews, P. S. Wheatley, P. B. Webb, P. Wormald and R. E. Morris, *Nature*, 2004, **430**, 1012.
50. Y.-P. Xu, Z.-J. Tian, S.-J. Wang, Y. Hu, L. Wang, B.-C. Wang, Y.-C. Ma, L. Hou, J.-Y. Yu and L.-W. Lin, *Angew Chem. Int. Ed.*, 2006, **45**, 3965.
51. R. Cai, M. Sun, Z. Chen, R. Munoz, C. O'Neill, D. E. Beving and Y. Yan, *Angew. Chem. Int. Ed.*, 2008, **47**, 525.
52. E. R. Parnham and R. E. Morris, *J. Am. Chem. Soc.*, 2006, **128**, 2204.
53. D. N. Dybtsev, H. Chun and K. Kim, *Chem. Commun.*, 2004, 1594.

CHAPTER 10

General Organic Reactions

10.1 INTRODUCTION

Over the past ten years, room-temperature ionic liquids have had a greater impact on organic synthesis than on other field of chemistry. The negligible vapour pressures, non-flammabilities, polarities, thermal stabilities, range of solubilities, coordination properties, and ease of product recovery and catalyst recycling make ionic liquids eminently suitable as reaction media for a wide variety of reactions in organic chemistry. In addition, rates of reaction and selectivities are often better in ionic liquids than in conventional molecular solvents.

The opportunity to select or design ionic liquids with chemical properties ideally suited to a specific reaction is an added attraction. For example, the low nucleophilicity of ionic liquid anions such as $[BF_4]^-$ and $[PF_6]^-$ provide unique environments for studying ionic reactions involving electron-deficient intermediates such as carbocations and onium ions.

The use of ionic liquids as reaction media and/or catalysts has now been investigated extensively for many types of general reactions in organic chemistry, *e.g.* hydrogenations, hydroformylations, oxidations, and also for numerous named reactions, such as Heck reactions, Suzuki reactions and Friedel–Crafts reactions (Chapter 11).

10.2 ALDOL CONDENSATION

The aldol reaction, also known as aldol addition, is an acid- or base-catalysed carbon–carbon bond forming reaction in which an enol or

An Introduction to Ionic Liquids
By Michael Freemantle
© Michael Freemantle 2010
Published by the Royal Society of Chemistry, www.rsc.org

enolate ion of a carbonyl compound reacts with another carbonyl compound to form a *β*-hydroxycarbonyl compound, the aldol. The reaction is an example of nucleophilic addition. The enol or enolate is the nucleophile, which can be an aldehyde or ketone, and the other carbonyl compound, which can also be an aldehyde or ketone, is its electrophilic partner. When the reaction occurs with the loss of water it is known as aldol condensation. The product of the reaction in this case is an *α,β*-unsaturated carbonyl compound. When two molecules of the same carbonyl compound condense, the reaction is known as self-aldol condensation. When the two molecules are different, the reaction is called cross-aldol condensation.

In 2002, a group led by Mehnert showed that imidazolium ionic liquids are suitable media for the self-aldol condensation of propanal (**10.1**) and the cross-aldol condensation of propanal and 2-methylpenta-nal (**10.2**).[1] The reactions were carried out using concentrated solutions of the catalyst sodium hydroxide in the ionic liquids. The products of the two reactions, 2-methylpent-2-enal (**10.3**) and 2,4-dimethylhept-2-enal (**10.4**) respectively (Scheme 10.1), were separated from the ionic liquid solution by distillation. Depending on the ionic liquid used, the selectivities for these aldehydes were significantly higher, compared with those using a sodium hydroxide–water catalyst system.

In a separate development, Medina and co-workers demonstrated that the ionic liquid choline hydroxide (**10.5**) can be used as a basic catalyst for the self- and cross-aldol condensation reactions of various aldehydes and ketones.[2] The researchers carried out the reactions using

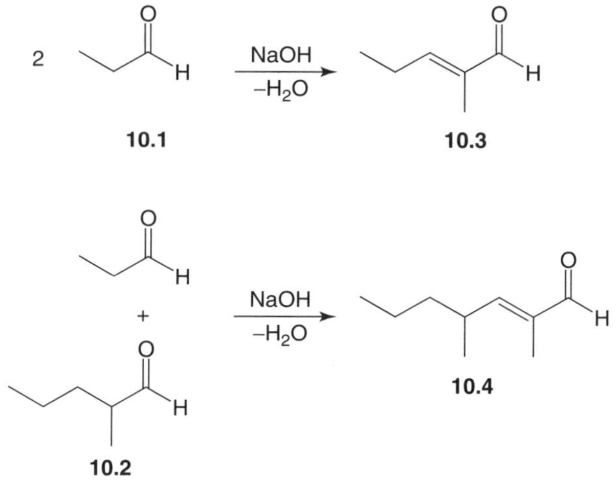

Scheme 10.1 Aldol condensations.

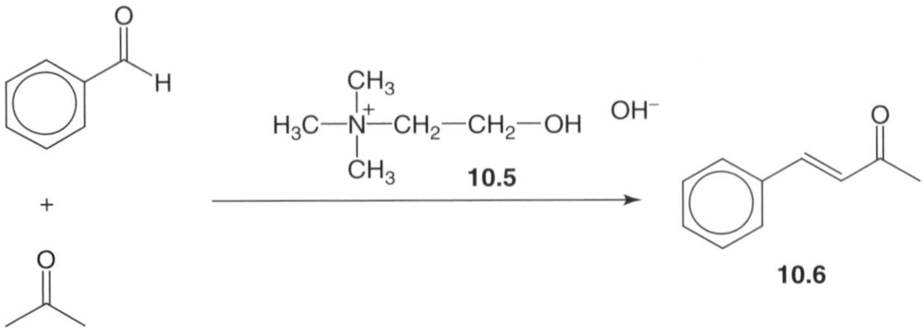

Scheme 10.2   Cross-aldol condensation catalysed by choline hydroxide.

a heterogeneous catalyst system prepared by impregnating a magnesium oxide support with the ionic liquid. The team compared the conversions, selectivities and turnover frequencies of the catalyst with those achieved using choline hydroxide without the support and magnesium oxide without the ionic liquid. In many cases the performance of the supported ionic liquid system was better. The aldol condensation of acetone and benzaldehyde, for example, resulted in higher conversion and selectivity for the product, benzylideneacetone (**10.6**), and also the highest turnover frequency (Scheme 10.2).

10.3   ALKYLATION

Ionic liquids can be employed as solvents and catalysts for alkylations using electrophilic alkylating agents. Examples are the Friedel–Crafts alkylations that involve the use a Lewis acid catalyst and alkyl halides as the alkylating agents (Section 11.5). Such alkylations have typically employed imidazolium chloroaluminate ionic liquids. Phosphonium ionic liquids have been used for alkylations with nucleophilic alkylating agents such as Grignard reagents (Section 11.6).

   Alkylations with electrophilic alkylating agents can also be carried out in non-chloroaluminate ionic liquids with imidazolium cations. In 1998, for example, Earle and co-workers described the use of [C$_4$mim][PF$_6$] for the base-catalysed regioselective alkylation of two compounds by nucleophilic displacement (Sections 8.7 and 10.17).[3] They used the ionic liquid as a "green" recyclable alternative to dipolar aprotic solvents, such as dimethyl sulfoxide and dimethylformamide, for the alkylation of the heteroatom of indole and 2-naphthol, both of which are nucleophiles. In these types of reactions, the nucleophiles displace the halide atoms in the alkyl halides that are used as the electrophilic alkylating agents (Scheme 8.2).

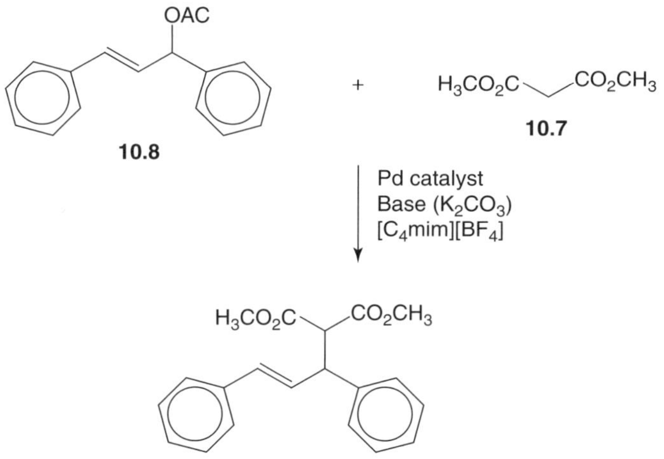

Scheme 10.3 Palladium-catalysed allylic alkylation

Another example of the alkylation of nucleophiles is the palladium-catalysed allylic alkylation of carbon nucleophiles in [C$_4$mim][BF$_4$].[4] Xiao and co-workers showed that, by varying the reaction conditions, it was possible to achieve up to 100% conversion for the alkylation of dimethyl malonate (**10.7**) by 3-acetoxy-1,3-diphenylprop-1-ene (**10.8**) (Scheme 10.3). The reaction proceeded readily with the benefit of easy recycling of the catalyst-containing ionic liquid phase. The product was extracted with toluene after each cycle.

Kabalka and co-workers have also studied transition-metal catalysed allylic alkylations in ionic liquids. They investigated the rhodium-catalysed cross-coupling reactions of allyl alcohols with a wide variety of boronic acids in [C$_4$mim][PF$_6$].[5] For example, they carried out the reaction of cinnamyl alcohol (**10.9**) and *p*-tolylboronic acid (**10.10**), Scheme 10.4). The yields varied immensely depending on reaction conditions and on the nature of the rhodium catalyst. The best yield, 72%, was achieved when a mixture of rhodium(III) chloride and copper(II) acetate was used as the catalyst.

10.4 CRACKING

Chloroaluminate ionic liquids have proved to be promising solvents and catalysts for cracking reactions. For example, Seddon and co-workers reported, in 2000, that polyethylene can be cracked in [C$_2$mim]Cl-AlCl$_3$ and related ionic liquids using an acidic co-catalyst such as concentrated sulfuric acid.[6]

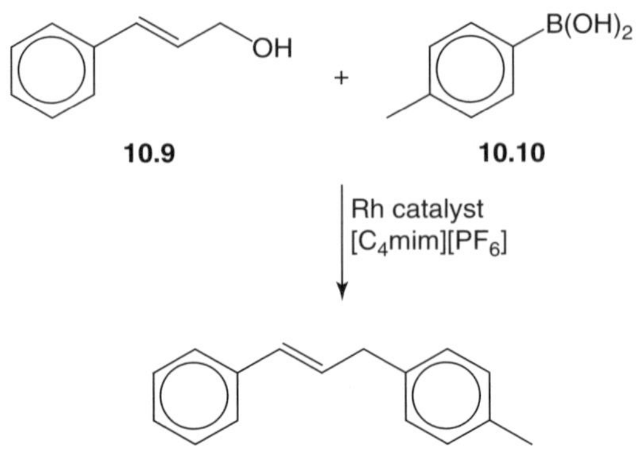

Scheme 10.4   Rhodium-catalysed allylic alkylation.

Polyethylene cracking is an important process for recycling plastics. Seddon's group showed that cracking the polymer in chloroaluminate ionic liquids achieved yields of up to 95% of a mixture of gaseous low molecular weight alkanes such as propane and isobutane as well as cyclic alkanes with low volatility. The distribution of the volatile alkanes depended on temperature. Unlike other polyethylene cracking processes, there was no significant formation of aromatic hydrocarbons or olefins. The products of the ionic liquid process were readily separated from the ionic liquid, allowing the ionic liquid to be recycled.

In 2007, Kamimura and Yamamoto demonstrated that ionic liquids can be used to convert 6-nylon (**10.11**), a polyamide, into good yields of the monomer caprolactam (**10.12**, Scheme 10.5).[7] They used *N,N*-dimethylaminopyridine as a catalyst. *N*-Methyl-*N*-propylpiperidinium and *N,N,N*-trimethyl-*N*-propylammonium salts with [NTf$_2$]$^-$ anions proved to be the most effective ionic liquids for the depolymerization. The monomer was collected by distillation and the ionic liquids were used repeatedly several times without decomposition, even though the depolymerization was carried out at 300 °C. The authors concluded that the work "will open a new field in ionic liquid chemistry as well as plastic recycling".

Li and colleagues reported that acidic [C$_4$mim]Cl-AlCl$_3$ is effective as a catalyst and solvent for the cracking of alkoxypropanes, such as 2,2-dimethoxypropane (**10.13**).[8] The cracking reactions at 130 °C produced good yields of alkoxypropenes and alcohols (Scheme 10.6) without the significant formation of esters that occurs with other alkoxypropane cracking processes. The use of [C$_4$mim][BF$_4$] resulted in a slower rate of reaction and much lower yields, indicating that the Lewis acidity of [C$_4$mim]Cl-AlCl$_3$ plays a critical role in the cracking process.

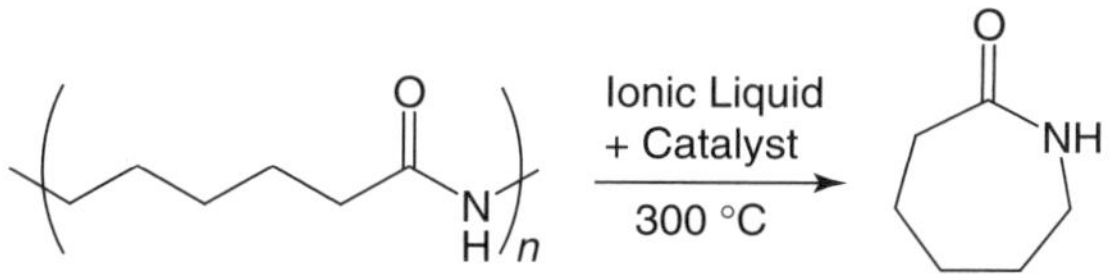

10.11   **10.12**

Scheme 10.5 Depolymerization of 6-nylon in an ionic liquid.

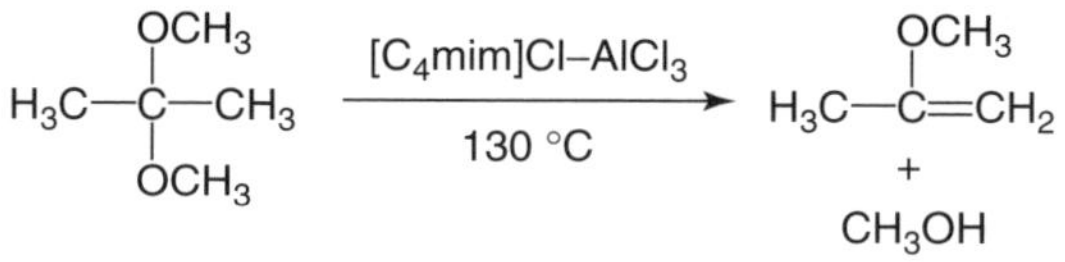

10.13

Scheme 10.6 Cracking of an alkoxypropane in an ionic liquid.

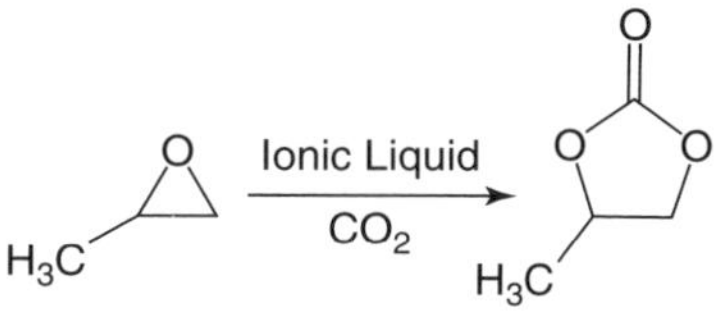

10.14   **10.15**

Scheme 10.7 Cycloaddition of CO_2 to propylene oxide.

10.5 CYCLOADDITION

Ionic liquids have been used for Diels–Alder reactions (Section 11.4) and other types of cycloaddition reactions.

One example is the use of ionic liquids with $[C_4mim]^+$ or $[C_4py]^+$ cations as catalysts for the cycloaddition of CO_2 to propylene oxide (**10.14**) to form propylene carbonate (**10.15**) (Scheme 10.7). Propylene carbonate is used as an aprotic polar solvent in organic synthesis and has a wide range of uses in the pharmaceutical, agrochemical and other industries.

Peng and Deng showed, in 2001, that $[C_4mim][BF_4]$ resulted in rapid cycloaddition of CO_2 to propylene oxide with almost 100% selectivity for the cyclic carbonate.[9] Furthermore, it was possible to separate the product by distillation, leaving the ionic liquid catalyst to be used again. The two chemists suggested that the $[C_4mim]^+$ cation provides an active catalytic site for the formation of a CO_2-propylene oxide complex during the reaction.

The following year, Deng and co-workers reported that imidazolium and pyridinium ionic liquids are suitable reaction media for the electrocatalytic cycloaddition of CO_2 to epoxides such as propylene oxide.[10] The cycloaddition was carried out without any additional supporting electrolyte and catalyst. In a typical experiment, an electrochemical cell with a copper cathode and a magnesium or aluminium anode was charged with the ionic liquid and the epoxide. CO_2 was then bubbled through a gas inlet. The product was extracted from the reaction mixture with diethyl ether.

In a further development, Kawanami and colleagues showed that propylene carbonate can be synthesized from propylene oxide using a supercritical CO_2/ionic liquid system.[11] They achieved almost 100% yield and 100% selectivity for the product within five minutes using the ionic liquid [C_8mim][BF_4] in supercritical CO_2. The turnover frequency was 77 times greater than those previously reported for the reaction.

10.6   DEBROMINATION

Ionic liquids can facilitate debrominations. For example, in 2001 Dyson and co-workers described the preparation of an ionic liquid with a transition metal carbonyl anion that catalyses the debromination of 2-bromoketones to the corresponding ketones.[12] The ionic liquid, [C_4mim][Co(CO)$_4$], was prepared by the metathesis reaction between [C_4mim]Cl and Na[Co(CO)$_4$]. It is a viscous liquid even at temperatures as low as 0 °C. A solution of NaOH in the ionic liquid was shown to catalyse the debromination of bromoacetophenone (**10.16**) in the absence of an organic solvent or other reagents (Scheme 10.8).

10.7   DIMERIZATION AND OLIGOMERIZATION

The use of ionic liquids and transition metal catalysts for the biphasic dimerization and oligomerization of linear alkenes to form dimers, trimers and other oligomers has been extensively investigated over the past few decades.

Early work on the topic was carried out by Chauvin and colleagues at the Institut Français du Pétrole. The group showed that

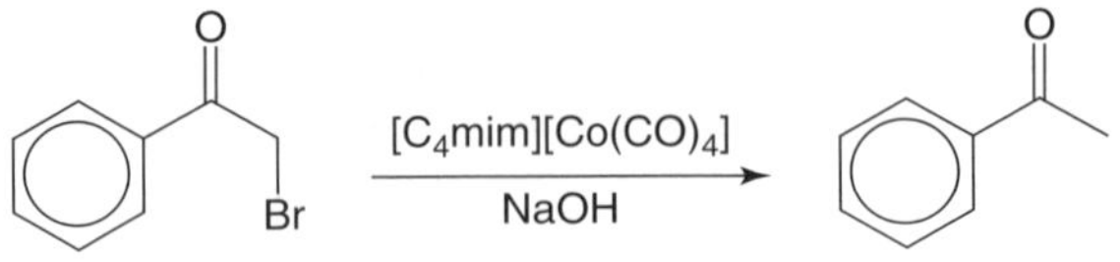

10.16

Scheme 10.8   Debromination.

chloroaluminate ionic liquids with $[C_4mim]^+$, $[C_4py]^+$ or $[P_{4\ 4\ 4\ 4}]^+$ cations can be used for the organonickel-catalysed dimerization of propene to form hexene isomers.[13] The group subsequently developed a continuous biphasic process, known as the Difasol process, for the catalytic dimerization of *n*-butene to isooctenes, which are used for the manufacture of isononanols (Section 14.3, the Difasol Process). The nickel catalyst is activated by alkylaluminium chloride compounds.

In 1999, Wasserscheid and co-workers reported the use of an organo-nickel catalyst (**10.17**) for the dimerization of but-1-ene that does not require an alkylaluminium compound for activation.[14] The biphasic catalytic dimerization was carried out in a slightly acidic $[C_4py]Cl\text{-}AlCl_3$ ionic liquid buffered with a weak organic base such as pyridine or pyr-role. The base suppresses the formation of free acid species in the ionic liquid. Such species can swamp the dimerization activity of the catalyst, resulting in the formation of branched higher oligomers. The process described by Wasserscheid's group proved to be selective for the linear C_8-olefin dimer with significant enhancement of catalytic activity com-pared with its activity in conventional organic solvents. The process also allowed easy product separation and catalyst recycling.

10.17

Wasserscheid, Gordon and co-workers subsequently showed that dialkylimidazolium ionic liquids, such as $[C_4mim][PF_6]$, with weakly coordinating anions can be used for the biphasic oligomerization of ethylene to higher alk-1-ene products using the cationic nickel complex **10.18** as a catalyst. The oligomerization proceeded with better reactivity and selectivity for the products than in conventional solvents.[15] Al-though the cationic nickel complex proved to be highly active in the ionic liquids, it was deactivated by traces of water.

10.18

Ionic liquids can also be used for the catalytic hydrodimerization of alkenes: a process that involves the reaction of alkenes with water to form dimers. In 1998, for example, Dupont and co-workers showed that palladium(II) compounds immobilized in [C$_4$mim][BF$_4$] catalyse the selective hydrodimerization of buta-1,3-diene (**10.19**) to form octa-1,3,6-triene (**10.20**) and octa-2,7-dien-1-ol (**10.21**) (Scheme 10.9).[16] The conversion of the diene and the turnover frequency of the catalyst increased significantly when the reaction was carried out under CO$_2$ pressure. The reaction was performed under homogeneous conditions at 70 °C. When the temperature was lowered to 5 °C, a two-phase system was formed, allowing the products to be easily separated from the reaction mixture by decantation. The catalyst-containing ionic liquid solution was reused several times without significant loss of catalytic activity.

In related work, the same group showed that buta-1,3-diene undergoes catalytic cyclodimerization to give 4-vinylcyclohexene (**10.22**) with 100% selectivity in [C$_4$mim][BF$_4$] (Scheme 10.10).[17] The catalyst for the biphasic process was an iron complex prepared by the reduction of the iron nitrosyl complex [Fe(NO)$_2$Cl]$_2$ *in situ* with a reducing agent in the ionic liquid. The product was separated from the reaction mixture by decantation. The ionic liquid/catalyst solution was reused several times without loss of catalytic activity or selectivity.

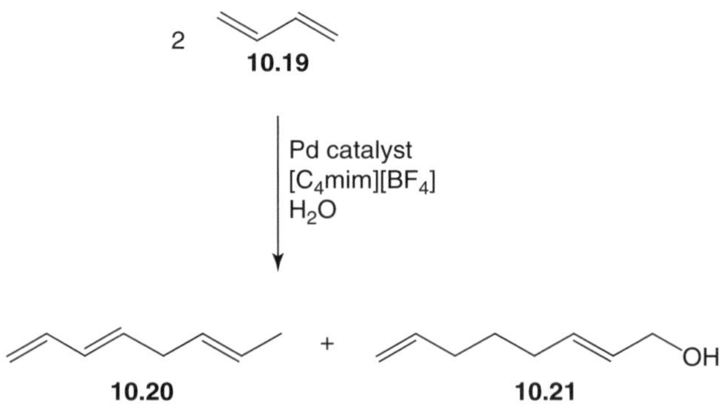

Scheme 10.9 Hydrodimerization.

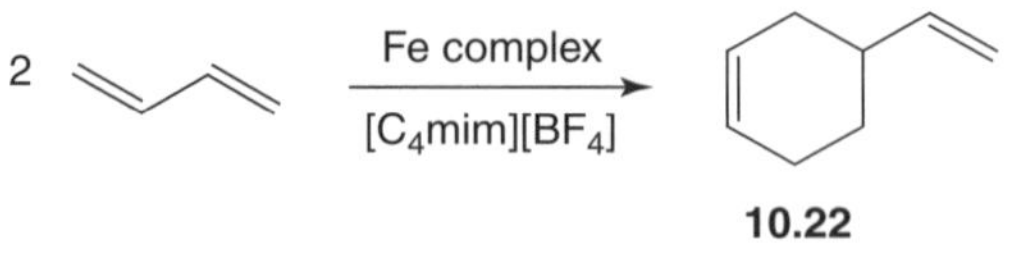

Scheme 10.10 Cyclodimerization.

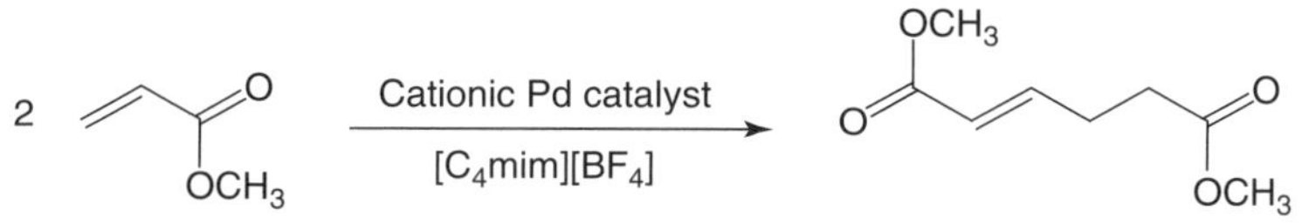

Scheme 10.11 Dimerization of methyl acrylate.

Ionic liquids can also be used for catalytic dimerization of functionalized olefins. In 2002, Wasserscheid and co-workers described the first continuous, biphasic, palladium-catalysed dimerization of methyl acrylate (**10.23**) (Scheme 10.11).[18] The group used a cationic ammonium phosphine ligand to immobilize the palladium catalyst in the ionic liquid, [C$_4$mim][BF$_4$]. The reaction was carried out continuously for 50 hours with an overall turnover number of more than 4000 and selectivity for the product, dimethyl Δ^2-dihydromuconate (**10.24**), of over 90%.

10.8 ELECTROPHILIC REACTIONS

Electrophiles are electron-deficient chemical species that accept electrons. They include Lewis acids and positive ions. Examples of electrophilic reactions that have been carried out in ionic liquids are halogenations aromatic nitrations and Friedel–Crafts alkylations and acylations (Sections 10.12, 10.16 and 11.5).

10.9 EPOXIDATION

In 2000, Song and Roh reported the use of an ionic liquid for epoxidation.[19] They developed a procedure for immobilizing and recycling Jacobsen's chiral manganese(III) salen epoxidation catalyst in [C$_4$mim][PF$_6$] and showed that the ionic liquid/catalyst system could be used repeatedly for the asymmetric epoxidation of alkenes (Section 8.9).

Three years later, Abu-Omar and colleagues reported that imidazolium and pyridinium ionic liquids can be used with the catalyst methyltrioxorhenium, CH$_3$ReO$_3$, for the epoxidation of alkenes and allylic alcohols.[20] They used urea hydrogen peroxide, (NH$_2$)$_2$CO·H$_2$O$_2$, a crystalline solid, as a water-free source of the oxidant hydrogen peroxide to yield epoxides such as **10.25**. The solid is soluble in ionic liquids like [C$_2$mim][BF$_4$] but only sparingly soluble in aqueous and organic media. The use of solutions of H$_2$O$_2$ and water for oxidations leads to the formation of diols such as **10.26** (Scheme 10.12).

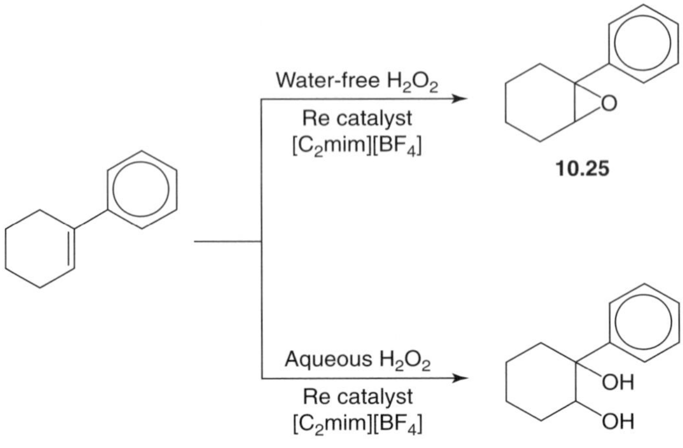

Scheme 10.12   Epoxidation requires water-free H$_2$O$_2$.

As with other organic reactions, the ionic liquids are only fully effective for epoxidations if they are free of halide impurities. Halides cause the hydrogen peroxide to decompose into oxygen and water, resulting in lower yields of the epoxides.

Chan and co-workers have tested the use of a manganese sulfate/sodium bicarbonate catalyst system for the epoxidation of alkenes in the water-miscible ionic liquid [C$_4$mim][BF$_4$] using aqueous H$_2$O$_2$ as the oxidant.[21] Although the catalyst is known to promote alkene epoxidation with aqueous H$_2$O$_2$, epoxidation using the ionic liquid as a solvent did not occur. The team attributed the lack of reaction to the poor solubility of sodium bicarbonate in [C$_4$mim][BF$_4$]. On replacing the sodium bicarbonate with tetramethylammonium hydrogencarbonate, the reaction proceeded readily to give yields of over 99% of some of the epoxides. The products were extracted with pentane and the ionic liquid reused repeatedly with good yields on addition of further small amounts of the catalyst system.

In 2004, Bernini and co-workers described the use of [C$_4$mim][BF$_4$] as a solvent for the epoxidation of 20 compounds with structures that occur in natural products.[22] The compounds included aromatic ketones called chalcones, *e.g.* **10.27** (Scheme 10.13). The procedure employed alkaline H$_2$O$_2$ as the oxidant. All the reactions resulted in yields of over 98% and were faster than in conventional solvents such as acetone. The epoxides were extracted from the reaction mixtures with diethyl ether.

10.27

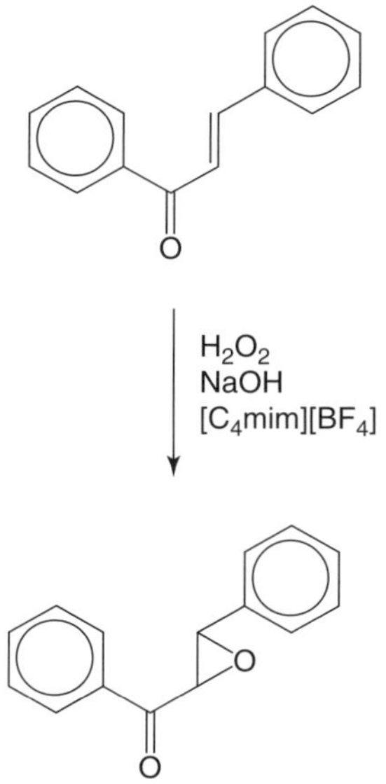

H$_2$O$_2$
NaOH
[C$_4$mim][BF$_4$]

Scheme 10.13 Chalcone expoxidation.

10.10 ESTERIFICATION

Brønsted acidic and neutral ionic liquids have proved suitable solvents for a broad range of esterification reactions. For example, Tang, He and colleagues demonstrated that carboxylic acids, such as acetic acid, could be esterified with alcohols, such as 1-butanol, in the Brønsted acidic ionic liquid [Hmim][BF$_4$].[23] The ionic liquid was prepared by mixing 1-methylimidazole with an aqueous solution of tetrafluoroboric acid, HBF$_4$, at 0 °C, followed by removal of the water under vacuum. Conversions of the acids or alcohols of over 97% were achieved with 100% selectivity for the ester.

A team led by Imrie synthesized ferrocenyl esters in [C$_4$mim][BF$_4$] and [C$_4$mim][PF$_6$] at room temperature using what is known as the DCC/DMAP protocol.[24] The protocol employs 1,3-dicyclohexyl-carbodiimide (DCC) as a coupling reagent and 4-dimethylaminopyridine (DMAP) as a nucleophilic catalyst. Yields of the phenolic esters of up to 100% were obtained, for example, from the reaction of ferrocenemonocarboxylic acid (**10.28**) and 4-chlorophenol (**10.29**) (Scheme 10.14).

The authors also reported "very high" yields and "efficient" recycling of the ionic liquid solvent. In some of the reactions, *N,N*'-dicyclohexylurea was formed as a by-product in the ionic liquid. Although the by-product could not be removed from the ionic liquid by ether extraction,

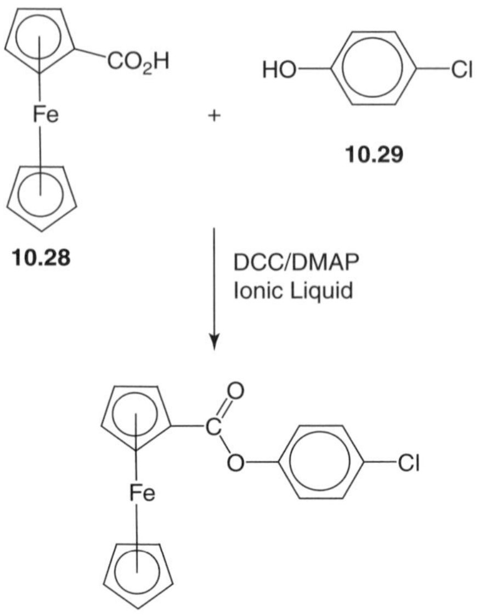

Scheme 10.14   Synthesis of a ferrocenyl ester.

it did not reduce the yield of the ester significantly over several recycles of the ionic liquid.

Srinivasan and co-workers achieved "excellent isolated yields" of esters from the acetylation of alcohols with acetic anhydride in various ionic liquids.[25] The reactions were carried out in dibutylimidazolium, $[C_4bim]^+$, ionic liquids at ambient temperature under ultrasound irradiation without the addition of a catalyst. The sonochemical *O*-acetylation of benzyl alcohol in $[C_4bim]Br$, for example, resulted in a 95% yield of benzyl acetate in five minutes.

In 2008, Williams, Welton and co-workers described an investigation of the influence of solvent basicity on acid-catalysed esterification in ionic liquids and molecular solvents such as toluene.[26] They studied the impact of three solvent dependent properties – hydrogen bond acidity, hydrogen bond basicity and dipolarity/polarizability – on the second-order rate constants for the esterification of benzyl alcohol (**10.30**) with methoxyacetic acid (**10.31**) (Scheme 10.15).

They concluded that the solvent's hydrogen bond basicity is the dominant influence on esterification rate. The best rates occurred in solvents with low hydrogen bond basicity. For example, the rate constant for the reaction in toluene was over twice that in $[C_4mim][NTf_2]$, which has a hydrogen bond basicity value more than three times that of toluene. The team explained that the solvent's hydrogen bond basicity determines the availability of the protons that catalyse the reaction. In

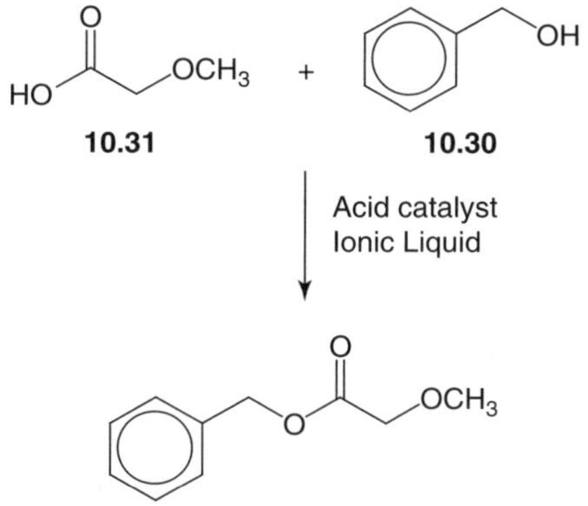

Scheme 10.15 Esterification of benzyl alcohol.

the case of ionic liquids, the hydrogen bond basicity is controlled by the anion. For example, the hydrogen bond basicity value of [C$_4$mim][OTf] is 0.51 compared with 0.21 for [C$_4$mim][NTf$_2$]. The second-order rate constant for the esterification in [C$_4$mim][NTf$_2$] is more than six times higher than that in [C$_4$mim][OTf].

10.11 ETHERIZATION

There has been some interest in the synthesis of ethers in ionic liquids. Asymmetrical tertiary alkyl ethers, such methyl *tert*-butyl ether (MTBE) (**10.32**), are used to boost the performance of gasoline. They are manufactured by the reactions of alkenes with alcohols. For example, MTBE is synthesized by the reaction of isobutene with methanol.

According to Deng and co-workers, there is a "growing problem" with the availability of isobutene as a raw material.[27] A possible alternative route to MTBE is the acid-catalysed etherization of *tert*-butyl alcohol with methanol (Scheme 10.16). The reaction leads to the loss of a molecule of water.

Deng's group tested various ionic liquids as non-acid catalysts and dehydrating agents for the synthesis of MTBE. The best results were

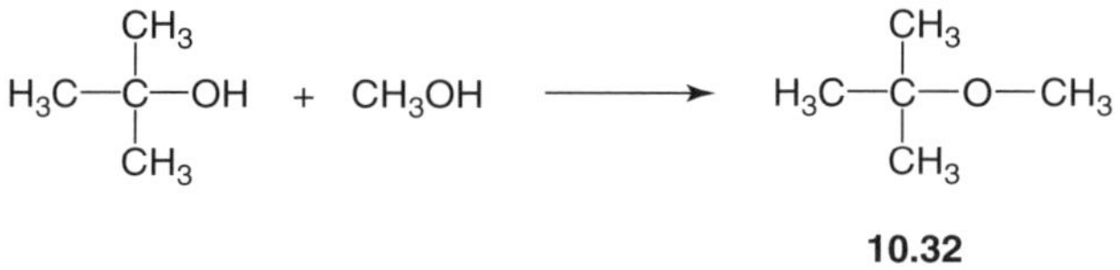

Scheme 10.16 Etherization.

obtained using [C$_{10}$mim][BF$_4$]. At 130 °C, they achieved a *tert*-butyl alcohol conversion of 93% and selectivity for MTBE of 97%. At 175 °C, the conversion rose to nearly 100% although the selectivity dropped to 80% owing to the formation of isobutene. The reaction mechanism of the process is not clear, according to the authors.

10.12 HALOGENATION

Compared with other types of organic reaction, relatively few studies have carried out on halogenations in ionic liquids. One of these studies, published in 2001, showed that imidazolium ionic liquids like [C$_4$mim][PF$_6$] might be suitable as "green" recyclable alternatives to chlorinated solvents, such as CCl$_4$, for the electrophilic addition of halogens to alkenes and alkynes.[28]

In another paper, Srinivasan and colleagues reported that aromatic substrates can be monobrominated regioselectively with *N*-bromo-succinimide (**10.33**) in the ionic liquid 1,3-dibutylimidazolium tetra-fluoroborate (**10.34**) without the use of a catalyst.[29] They described the isolated yields of the products as "excellent". For example, they achieved a 96% yield of 4-bromoacetanilide (**10.35**) from the bromination of acetanilide (**10.36**) in the ionic liquid (Scheme 10.17).

In 2004, Earle and co-workers reported the oxidative mono-halogenation of benzene, toluene and anisole, C$_6$H$_5$OCH$_3$, with HCl or

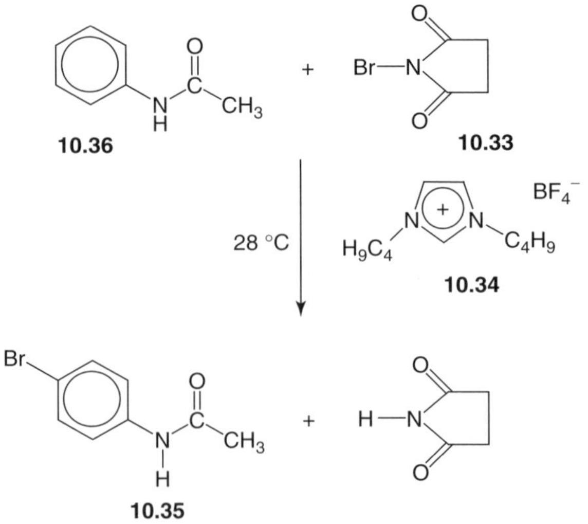

Scheme 10.17   Bromination.

HBr using the ionic liquid [C₄mim][NO₃] as a catalyst and air as an oxidant.[30] During the reaction, the nitrate cation was reduced to nitrous oxide, N_2O, and the HCl or HBr oxidized to the hypohalous acids, HOCl or HOBr respectively. These acids are well-known halogenating agents. The N_2O, which is unstable in the acidic conditions, was oxidized back to nitrate by air. Earle's team also showed that the reaction of toluene with nitric acid in halide ionic liquids yielded halotoluenes.

10.13 HYDROFORMYLATION

Hydroformylation is the reaction of carbon monoxide, hydrogen and an alkene to yield an aldehyde. A transition metal complex, typically a rhodium complex with triphenylphosphine ligands, is used to catalyse the reaction.

In 1972, Parshall described the use of tetraethylammonium ionic liquids for the platinum-catalysed hydroformylation of ethene (Section 8.2).[31] Knifton subsequently published several papers and filed several patents focusing on the use of tetraalkylphosphonium and tetraalkylammonium ionic liquids, such as [P₄ ₄ ₄ ₄]Br and [N₄ ₄ ₄ ₄]Br, respectively, for the preparation of aldehydes and other compounds using syngas (synthesis gas, $CO + H_2$). He demonstrated, for example, that aldehydes can be prepared by the hydroformylation of straight-chain terminal olefins, such as 1-butene, and linear internal olefins, such as 2-octene, in various ionic liquids using various ruthenium catalysts.[32]

Chauvin's group published a report on the rhodium-catalysed biphasic hydroformylation of pent-1-ene in [C₄mim][PF₆] in 1995.[33] The hexanal products proved to be poorly soluble in the ionic liquid. Triphenylphosphine ligands were used to immobilize the rhodium catalyst precursor in the ionic liquid and avoid the catalyst leaching into the organic product phase.

Supported ionic liquid phase (SILP) catalysis (Section 8.5) has been used for rhodium-catalysed hydroformylation of 1-hexene to heptanal.[34] The rhodium phosphine catalyst was dissolved in a layer of [C₄mim][PF₆] or [C₄mim][BF₄], the cations of which were covalently anchored to the surface of a silica gel support. The use of [C₄mim][BF₄] was limited by the aldehyde concentrations in the reaction mixture. At high concentrations, the ionic liquid partially dissolved in the organic phase, resulting in leaching of the rhodium catalyst.

Agbossou-Niedercorn and co-workers showed that the biphasic hydroformylation of olefins can be achieved at "commercially competitive rates" in imidazolium triflate ionic liquids.[35] The ionic liquid

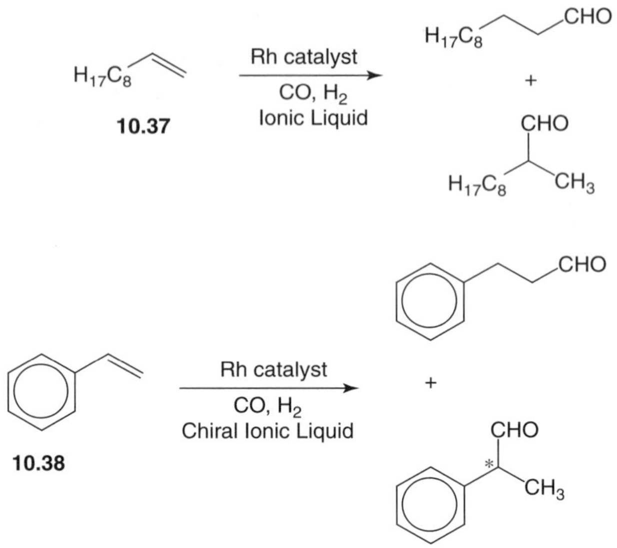

Scheme 10.18   Hydroformylation of dec-1-ene and styrene.

phase contained the catalyst: a rhodium complex modified with tri-phenylphosphine ligands. The hydroformylation of dec-1-ene (**10.37**) using this catalyst yielded a mixture of linear and branched aldehydes (Scheme 10.18). The conversions and ratio of linear to branched alde-hydes depended on the ionic liquid.

The group also investigated the use of the chiral ionic liquid, 1-(2-methylbutyl)-3-methylimidazolium triflate for the rhodium-catalysed hydroformylation of styrene (**10.38**). The reaction proceeded with a small amount of enantioselectivity. The groups suggested that the chiral imidazolium cation creates a chiral environment around the active rhodium centre and therefore transmits chiral information.

10.14   HYDROGENATION

The transition metal catalysed hydrogenation of olefins, arenes and other compounds in ionic liquids has been a subject of intense investi-gation since the mid-1990s. The earliest work was carried by Chauvin and colleagues in 1995. They showed that ionic liquids with weakly coordinating anions such as $[SbF_6]^-$ can be employed as solvents for the biphasic hydrogenation of pent-1-ene using a rhodium complex as catalyst.[33] The yield of the product pentane in $[C_4mim][SbF_6]$ was over twice that obtained in acetone and the turnover frequency around five times higher. They suggested that the enhanced performance resulted from stabilization of an active cationic rhodium(III) in the ionic liquid.

The group also carried out the asymmetric biphasic hydrogenation of α-acetamidocinnamic acid using the same ionic liquid and a chiral cationic rhodium complex as catalyst (Section 8.9).

The following year, Dupont and co-workers reported that $[C_4mim][BF_4]$, $[C_4mim][PF_6]$ and $[C_4mim]Cl\text{-}AlCl_3$ are effective solvents for the biphasic rhodium-catalysed hydrogenation of cyclohexene.[36] They found, for example, that the triphenylphosphine rhodium(I) catalyst, known as Wilkinson's catalyst, $RhCl(PPh_3)_3$, produces a stable solution in all three ionic liquids, from which they cannot be removed by non-polar organic solvents.

In 1997, the same group described the use of a biphasic $[C_4mim][BF_4]/$ isopropanol system for the ruthenium-catalysed asymmetric hydrogenation of 2-arylacrylic acids.[37] The team used a rhodium catalyst precursor with chiral ligands known as "BINAP" ligands [2,2′-bis(diphenylphosphino)-1,1′-binaphthyl]. The asymmetric hydrogenation of 2-(6-methoxy-2-naphthyl)acrylic acid (10.39) using a Ru-(S)-BINAP catalyst, for example, yielded the anti-inflammatory drug (S)-naproxen (10.40) quantitatively in 80% enantiomeric excess (Scheme 10.19). The product-containing isopropanol phase was separated by decantation and the catalyst-containing ionic liquid was recycled several times without "significant changes in activity and selectivity".

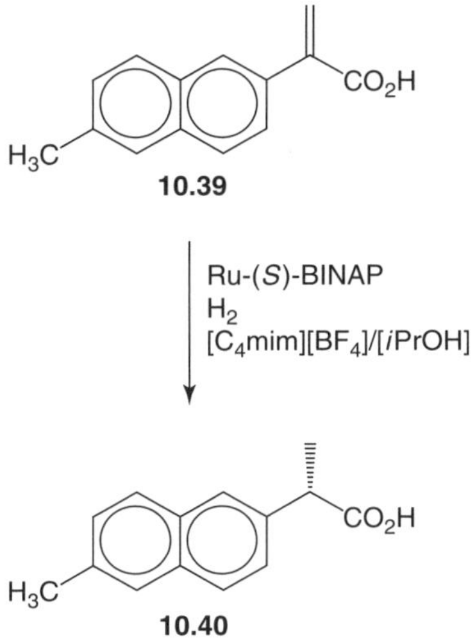

10.39

Ru-(S)-BINAP
H_2
$[C_4mim][BF_4]/[iPrOH]$

10.40

Scheme 10.19 Asymmetric hydrogenation.

Two years later, Dyson and co-workers reported the ruthenium-catalysed biphasic hydrogenation of arenes such as benzene, toluene and cumene, $C_6H_5CH(CH_3)_2$, in [C4mim][BF4].[38] The catalyst precursor was a cationic complex containing a cluster of four ruthenium atoms. The conversions and turnover frequencies in the ionic liquid using the cluster catalyst were comparable to those obtained using water as the solvent.

In a more recent development, Livingston and co-workers used a technique known as organic solvent nanofiltration to separate hydrogenation products from ionic liquid phases containing Ru-BINAP catalysts.[39] The technique employs a polyimide membrane that retains the catalyst and ionic liquid whilst allowing the permeation of the product. Livingston's group tested the system for the asymmetric hydrogenation of dimethyl itaconate (**10.41**) to dimethyl methylsuccinate (**10.42**) in [P6 6 6 14]Cl using the Ru-BINAP catalyst **10.43** (Scheme 10.20). They reported that the technique not only enhanced enantioselectivity but also improved the stability of the Ru-BINAP catalyst.

The use of ionic liquids as solvents for nanoparticle hydrogenation catalysts came under scrutiny in the early 2000s (Section 8.6). In 2002 for example, Dupont's group demonstrated that [C4mim][PF6] is not only an effective medium for the preparation and stabilization of iridium(0) nanoparticles but also that the nanoparticle catalyst/ionic liquid solution can be used for biphasic hydrogenations of olefins such as 1-decene and cyclohexene.[40]

Following on from that work, Dupont's group showed that ruthenium(0) nanoparticles can be generated in [C4mim][PF6] and related ionic liquids.[41] The group used the nanoparticle catalyst/ionic liquid solutions for the partial hydrogenation of benzene to cyclohexene.

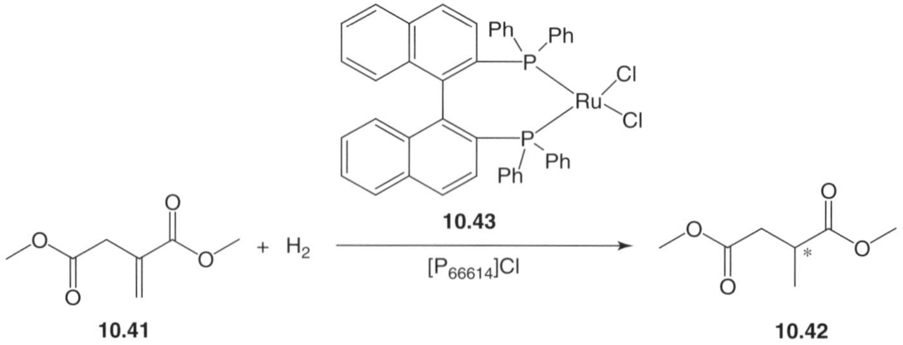

Scheme 10.20 Hydrogenation of dimethyl itaconate with a ruthenium-BINAP catalyst.

Supercritical CO_2 has featured significantly in biphasic hydrogenations that employ an ionic liquid as the other phase. Jessop and co-workers, for example, hydrogenated tiglic acid, 2-methylbut-2-enoic acid (**10.44**), using a ruthenium-BINAP catalyst in such a system (Scheme 10.21).[42] [C_4mim][PF_6] with some added water was used as the catalyst-containing ionic liquid phase. The supercritical CO_2 was used to extract the product, 2-methylbutanoic acid (**10.45**), leaving the catalyst/ionic liquid solution, which was reused repeatedly without significant loss of yield or enantioselectivity. Esters of the product are used as antihypercholesterolemic and vasodilator drugs.

The same group also used the procedure with the same catalyst and ionic liquid to carry out the asymmetric hydrogenation of iso-butylatropic acid (**10.46**) to form the anti-inflammatory drug ibuprofen (**10.47**) (Scheme 10.22). However, whereas water is an effective co-solvent with the ionic liquid for the hydrogenation of tiglic acid, the enantioselectivity for the asymmetric hydrogenation of isobutylatropic acid proved to be poor in the wet ionic liquid. The addition of methanol to the ionic liquid resulted in much higher enantioselectivity – higher even than the enantioselectivities reported for the reaction in aqueous/organic biphasic systems using the same catalyst.

In same year, Baker, Tumas and co-workers described biphasic hydrogenations using Wilkinson's catalyst immobilized in [C_4mim][PF_6] as one phase and supercritical CO_2 as the product-extracting phase.[43]

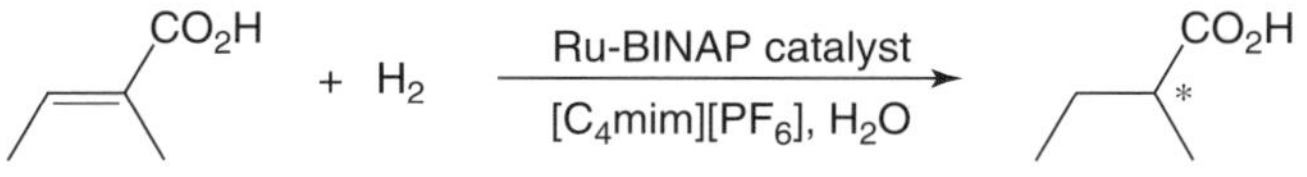

Scheme 10.21 Asymmetric hydrogenation of tiglic acid.

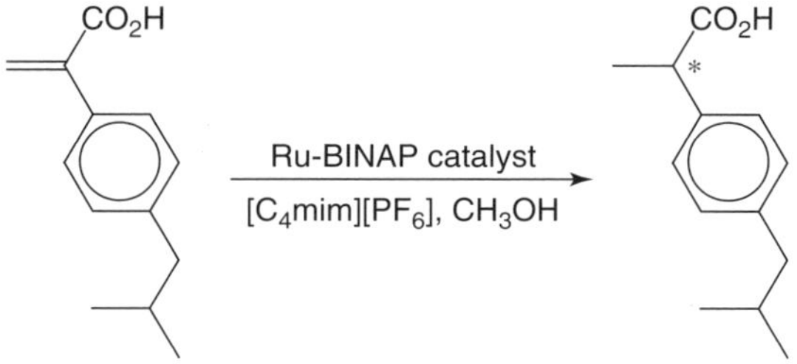

Scheme 10.22 Asymmetric hydrogenation yields ibuprofen.

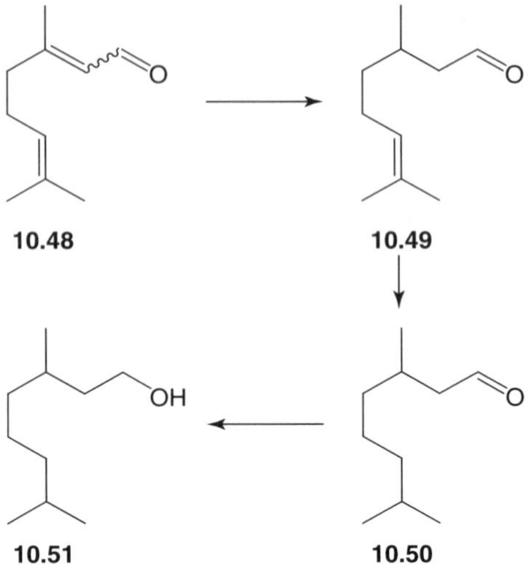

Scheme 10.23 Hydrogenation of citral.

They showed that the supercritical CO_2/ionic liquid system offers no reactivity advantage for the hydrogenation of alkenes, such as dec-1-ene or cyclohexene, compared with a biphasic system using the same ionic liquid and *n*-hexane as the other phase. They also demonstrated that the hydrogenation of CO_2 in the presence of dialkylamines proceeded with high selectivity for the *N,N*-dialkylformamide products. The group attributed the high selectivity to the formation of ionic carbamate intermediates in the ionic liquid.

Ionic liquids can also promote the product selectivity of supported hydrogenation catalysts. In 2008, Claus and co-workers reported that palladium catalysts supported on carbon or silica and coated with an ionic liquid are highly active and selective for the hydrogenation of citral, a lemon-scented compound that is used in the perfume industry (Scheme 10.23).[44] Citral (**10.48**) is a mixture of two geometric isomers: *trans*-citral, known as geranial, and *cis*-citral, known as neral. Using a Pd/C catalyst coated with the ionic liquid [C$_4$mim][dca], the group achieved 100% conversion of citral and over 99% selectivity for the product citronellal (**10.49**), an insect repellent, without further hydrogenation to the dihydro- and tetrahydro-products **10.50** and **10.51**.

10.15 HYDROSILYLATION

Hydrosilylation is a method of preparing organosilicon compounds by the addition of a silane group, Si–H, to unsaturated bonds such as carbon–carbon double bonds in alkenes. Several studies have shown that catalytic hydrosilylation of alkenes with silane-functionalized polysiloxanes is possible in ionic liquids.

Siloxanes are compounds that contain the Si–O–Si linkage. Polysiloxanes, $(R_2SiO)_n$, consist of an inorganic backbone, -O-Si-O-Si-O-Si- with organic groups, R, attached to the silicon atoms. In silane-functionalized polysiloxanes some of the organic functions (Si–R) are replaced by silane (Si–H) functions.

In 2006, Dyson and co-workers described a study of the biphasic platinum-catalysed hydrosilylation of 1-hexadecene (**10.52**) with polydimethyl-methylhydrogen-polysiloxane (**10.53**) in ionic liquids with pyridinium, piccolinium and imidazolium cations and anions such as $[BF_4]^-$ and $[NTf_2]^-$ (Scheme 10.24).[45] They used K_2PtCl_4 or $Pt(PPh_3)_4$ as catalyst precursors. The product was removed at the end of the reaction by decantation or with a syringe and the ionic liquid/catalyst system reused.

The performance of the ionic liquid/catalyst system in terms of catalyst activity, stability and recyclability varied, depending on the nature of the ionic liquid and catalyst. For example, a system consisting of $[C_4py][BF_4]$ and K_2PCl_4 resulted in a conversion of 98% whereas that for $[C_4mim][BF_4]$ and $Pt(PPh_3)_4$ was 75%. The purity of the ionic liquid

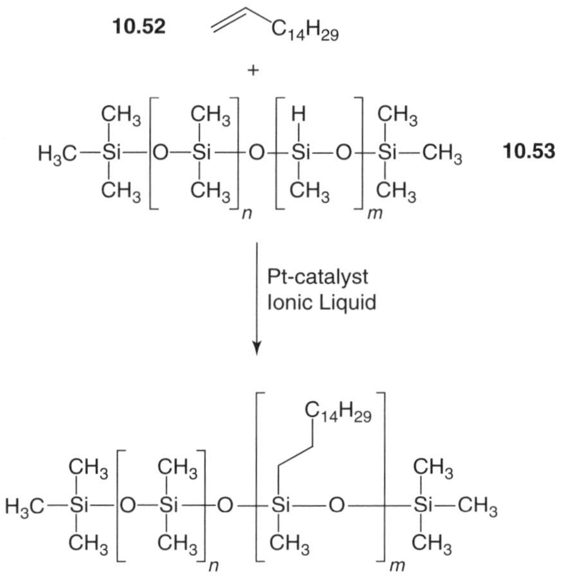

Scheme 10.24 Hydrosilylation.

also proved important as the catalyst immobilized in the ionic liquid is particularly sensitive to oxygen, water and halide impurities.

10.16 NITRATION OF AROMATICS

Aromatic nitrations are electrophilic aromatic substitution reactions. The electrophile, a nitronium ion, $[NO_2]^+$, which reacts with the benzene ring by removing an electron, is typically generated by the reaction of concentrated nitric acid with concentrated sulfuric acid which acts as a catalyst. Aromatic nitrations in industry employ large quantities of these concentrated acids. However, the extraction of nitrated aromatics from the acids, the regeneration and recycling of the acids, and the disposal of large amounts of environmentally harmful acidic and water-reactive wastes are major technological challenges.

The potential use of ionic liquids as solvents to generate milder and more environmentally compatible nitrating agents and facilitate product isolation and recovery has attracted the interest of chemists for many years. In 2001, for example, Laali and Gettwert reported a study of the electrophilic nitration of aromatic substrates using several nitrating agents in various ionic liquids.[46]

They showed that the selection of both the nitrating agent and the ionic liquid was critical. For example, the use of nitronium tetra-fluoroborate, $[NO_2][BF_4]$, as a nitrating agent and $[C_2mim][BF_4]$ as a solvent resulted in nitration of the imidazole ring of the ionic liquid's $[C_2mim]^+$ cation rather than nitration of the aromatic substrate. The best results, in terms of nitration efficiency of the aromatic compounds and recycling of the ionic liquid, were achieved using a combination of ammonium nitrate and trifluoroacetic anhydride as the nitrating agent and the ionic liquid $[C_2mim][CF_3CO_2]$ as the solvent.

The following year, Handy and Egrie reported the use of 70% aqueous nitric acid and ytterbium triflate, $Yb(OTf)_3$, as a catalyst to nitrate various aromatic substrates using the room-temperature ionic liquid 1-butyl-1-methylpyrrolidinium bistriflamide as a solvent.[47] The nitration of toluene, for example, resulted in a 95% yield of the three nitrobenzene isomers. Similar results were obtained using the less expensive catalyst $Cu(OTf)_2$. In both cases, the products were separated by distillation and the ionic liquid and catalyst re-used several times.

In 2003, Lancaster and Llopis-Mestre demonstrated that aromatic compounds such as toluene can be nitrated in high yields in ionic liquids without the use of a metal catalyst.[48] Instead, they used acetyl nitrate, which is a fairly mild nitrating agent. The best results were obtained

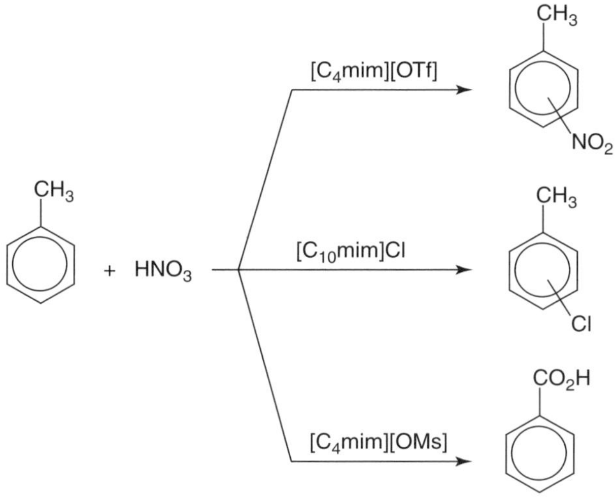

Scheme 10.25 Ionic liquids influence the outcome of a reaction.

using the ionic liquid *N*-butyl-*N*-methypyrrolidinium bistriflamide. The acetyl nitrate, which is formed *in situ* by the reaction of nitric acid and acetic anhydride, completely dissociates in the ionic liquid to acetate and the nitrating agent nitronium, $[NO_2]^+$.

The reaction of toluene and nitric acid provided the first example of how different ionic liquids produce different reaction types and products.[30] Earle and co-workers showed that the nitration of toluene with 67% nitric acid using [C$_4$mim][OTf] as a nitration catalyst was rapid and exothermic. The nitration resulted in a 99% yield of nitrotoluene (**10.54**) with a 2 : 1 ratio of the *para-*:*ortho*-isomers (Scheme 10.25).

In contrast, the reaction of toluene and nitric acid using a halide ionic liquid like [C$_{10}$mim]Cl as a catalyst afforded monohalogenated products. When an ionic liquid with methanesulfonate [OMs]$^-$ anions, such as [C$_4$mim][OMs], was used as a catalyst, the nitric acid acted as an oxidizing agent. The product was benzoic acid (**10.55**). The results show that the choice of ionic liquid can dramatically influence the outcome of a chemical reaction.

10.17 NUCLEOPHILIC REACTIONS

Nucleophiles are chemical species capable of donating electrons. They include Lewis bases and many negative ions. In nucleophilic displacement reactions, also known as nucleophilic substitutions, a nucleophile displaces a group attached to a positively charged atom. Nucleophilic displacement is one of the most widely used reactions in organic chemistry.

An example of nucleophilic displacement in an ionic liquid as solvent is the based-catalysed regioselective alkylation of indole and 2-naphthol in [C$_4$mim][PF$_6$] (Section 10.3). Many other examples have been reported in the scientific literature.

In 2001, for instance, Eckert and co-workers described the nucleophilic displacement of the chloride group on benzyl chloride, C$_6$H$_5$CH$_2$Cl, by cyanide to give phenylacetonitrile, C$_6$H$_5$CH$_2$CN, in [C$_4$mim][PF$_6$].[49]

The following year, Ren and co-workers reported the use of halide ionic liquids such as [C$_4$mim]Cl as reaction media for the coupling of aryl halides with sodium cyanide, NaCN, to yield aryl nitriles.[50] They used copper(I) cyanide (CuCN) or copper(I) iodide as a catalyst and achieved up to 100% yields for the conversion of iodobenzene, 2-iodonaphthalene and other aryl halides into the corresponding aryl nitriles. The team reported that the copper catalysts immobilized in the ionic liquids can be reused continuously.

Shieh and co-workers tested various ionic liquids as solvents and additives for the benzylation reaction of imidazoles, such as benzimidazole (**10.56**), with dibenzyl carbonate (**10.57**) using DABCO (1,4-diazabicyclo[2.2.2]octane) (**10.58**) as a nucleophilic catalyst (Scheme 10.26).[51]

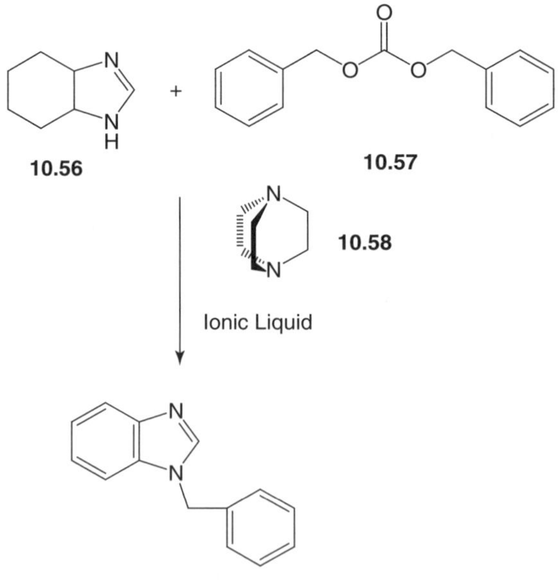

Scheme 10.26   Benzylation of benzimidazole.

Imidazolium and tetrabutylammonium ionic liquids with various anions all gave higher yields of the product compared with the reaction in acetonitrile. The team also observed significant rate enhancements for the reaction in the ionic liquid tetrabutylammonium chloride using microwave irradiation, with the reaction time being reduced from hours to minutes.

Lourenço and Afonso showed that [C$_4$mim][PF$_6$] is an efficient solvent and catalyst for promoting biphasic nucleophilic substitution reactions of alkyl halides using nucleophiles such as azide, cyanide, iodide and phenoxide.[52] They carried out the reactions under biphasic conditions using an aqueous solution of KOH or NaOH. Yields of up to 98% were achieved.

10.18 OLEFIN METATHESIS

Over the past decade or so, olefin metathesis catalysed by transition metal complexes, notably ruthenium carbene complexes, has become an important tool in organic synthesis. The importance of this type of reaction was recognized in 2005 with the award of the Nobel Prize in Chemistry to Chauvin, Grubbs and Schrock "for the development of the metathesis method in organic synthesis".

There are three categories of olefin metathesis. All involve the transposition of double bonds.

In the first, known as cross metathesis, the double bond of one olefin is transposed with the double bond of another olefin to form two other olefins. A variation of cross metathesis is self-cross metathesis, also known as self-metathesis, where two molecules of the same olefin react to generate one or more new olefins. For example, a compound with a terminal double bond, R-CH=CH$_2$, may combine with itself to form the product R-CH=CH-R. Another example is the self-metathesis of 2-pentene, CH$_3$CH=CHCH$_2$CH$_3$, to form 2-butene, CH$_3$CH=CHCH$_3$, and 3-hexene, CH$_3$CH$_2$CH=CHCH$_2$CH$_3$.

In the second type of olefin metathesis, called ring-closing metathesis, a diene with two terminal double bonds forms a cyclic olefin and another olefin.

Finally, ring-opening metathesis is the reverse of ring-closing metathesis. A cyclic olefin combines with an acyclic olefin to form a diene with two terminal double bonds.

The earliest work on olefin metathesis in ionic liquids was described briefly by Chauvin and Olivier-Bourbigou in 1995 (Section 8.8). They obtained "good results" for the metathesis of 2-pentene using a tungsten catalyst dissolved in Lewis acidic chloroaluminate ionic liquids.

Since then, several olefin self-metathesis reactions in ionic liquids have been carried out. Several studies have focused on the use of first- and second-generation Grubbs catalysts. First-generation Grubbs catalysts are ruthenium carbene complexes. The second generation are ruthenium catalysts with *N*-heterocyclic carbene ligands.

Williams and co-workers showed that the self-metathesis of 1-octene, $CH_3(CH_2)_5CH=CH_2$, to form 7-tetradecene, $CH_3(CH_2)_5CH=CH(CH_2)_5CH_3$, can be carried out in ionic liquids using a second-generation Grubbs catalyst (**10.59**).[53] The team achieved 98% conversion of 1-octene and 95% selectivity towards 7-tetradecene in [C$_4$mim][PF$_6$]. When the 1-octene self-metathesis reaction was carried out in the ionic liquid with a first-generation Grubbs catalyst, the catalyst rapidly became deactivated and leached into the product-containing organic layer.

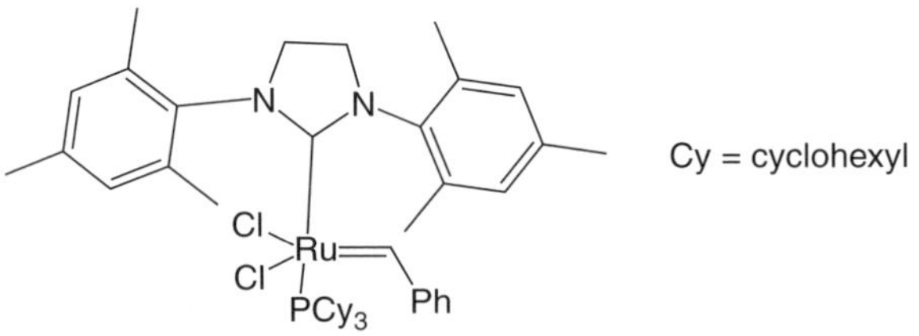

Cy = cyclohexyl

10.59

The use of first- and second-generation Grubbs catalysts for olefin ring-closing metathesis reactions in ionic liquids has also been extensively studied since the early 2000s. In 2001, Buijsman and co-workers established that [C$_4$mim][PF$_6$] is "an excellent solvent" in which to carry out ring-closing metathesis reactions of dienes using Grubbs catalysts.[54] They achieved, for example, 100% conversion of diene **10.60** using a first-generation Grubbs catalyst (**10.61**) (Scheme 10.27).

Mauduit, Guillemin and co-workers investigated the ring-closing metathesis of a similar diene in [C$_4$mim][PF$_6$] using a first-generation Grubbs catalyst tethered to an imidazolium cation.[55] The purpose of the

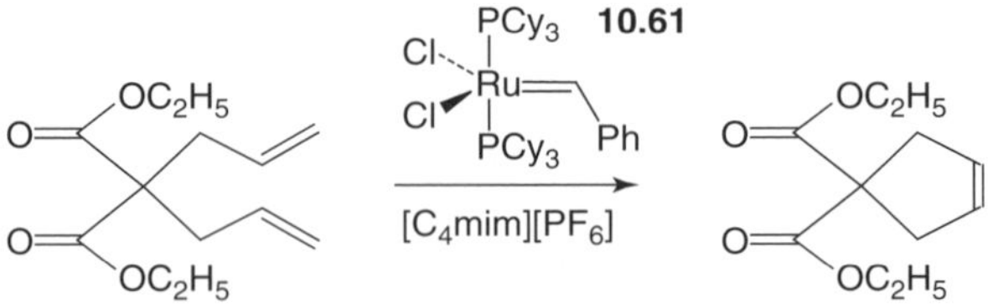

10.60

Scheme 10.27   Ring-closing metathesis.

cation tag was to avoid the catalyst leaching from the ionic liquid phase into the product-containing organic phase. The researchers observed that the ionic liquid-ruthenium catalyst exhibited high activity and "remarkable recyclability". They were also able to store it in [C$_4$mim][PF$_6$] for several months without loss of activity.

Olefin metathesis in ionic liquids using other types of transition metal catalysts has also been investigated. For example, in 2007 Miller II and Bowden concluded that [C$_4$mim][PF$_6$] is an "attractive new solvent" for olefin ring-closing and cross metathesis reactions using a molybdenum-based catalyst, known as Schrock's catalyst (**10.62**).[56] Conversions of over 97% were achieved with various olefin substrates.

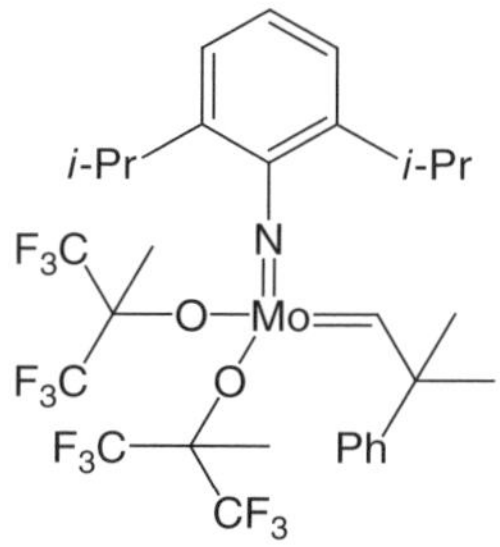

10.62

Dixneuf and co-workers used cationic ruthenium allenylidene salts such as **10.63** as catalysts for the ring-opening metathesis polymerization of norbornene (**10.64**) in an ionic liquid (Scheme 10.28).[57] The catalyst was dissolved in the ionic liquid 1-butyl-2,3-dimethylimidazolium hexa-fluorophosphate, [C$_4$dmim][PF$_6$], **10.65**. Good yields of the polymer **10.66** were obtained even after recycling the ionic liquid and catalyst several times. When Grubbs first- and second-generation catalysts were

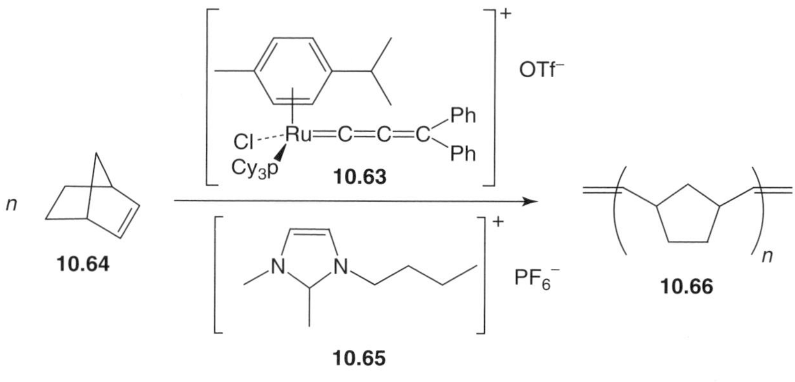

Scheme 10.28 Ring-opening metathesis polymerization of norbornene.

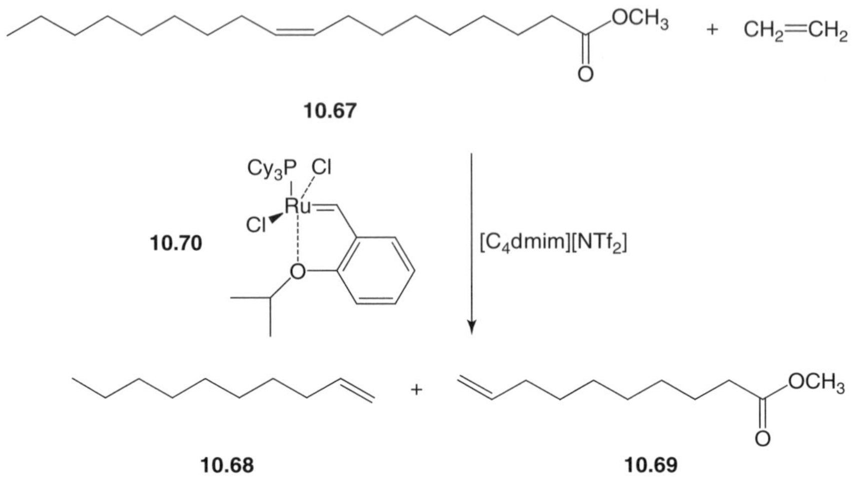

Scheme 10.29   Cross metathesis.

used in place of the ruthenium allenylidene salts, significant drops in catalytic activity were observed on recycling.

The same group also studied the ruthenium-catalysed cross-metathesis of methyl oleate (**10.67**) with ethylene in imidazolium ionic liquids (Scheme 10.29).[58] This type of reaction with ethylene is known as ethenolysis.

Methyl oleate, an unsaturated fatty ester derived from unsaturated vegetable oils, is a potentially attractive renewable feedstock for the chemical industry. The products of the olefin metathesis of methyl oleate in ionic liquids, 1-decene (**10.68**) and methyl 9-decenoate (**10.69**), are important chemical intermediates in the manufacture of products such as lubricants and polyesters.

Dixneuf's group achieved 83% conversion of methyl oleate with 100% selectivity for the 1-decene and methyl 9-decenoate in an ionic liquid with 1-butyl-2,3-dimethylimidazolium cations (**10.65**) [C_4dmim][NTf_2], using a type of ruthenium catalyst with tricyclohexylphosphine ligands known as a first-generation Hoveyda catalyst (**10.70**). The procedure required more catalyst than that used when the reaction was carried out in toluene. However, the catalyst/ionic liquid system could be recycled for three runs without loss of activity.

10.19   OXIDATION

Since the early 2000s, there have been numerous studies of the use of ionic liquids for the catalytic oxidations of organic compounds such as aldehydes, alcohols, alkanes and olefins.[59]

In 2000, Howarth described the oxidation of aromatic aldehydes in the ionic liquid [C$_4$mim][PF$_6$].[60] He used nickel(II) acetylacetonate, Ni(acac)$_2$, as catalyst and oxygen to oxidize seven aldehydes to the corresponding carboxylic acids. They included, for example, the oxidation of benzaldehyde to benzoic acid. After extraction of the acids with ethyl acetate, it was possible to reuse the catalyst/ionic liquid system several times without leaching of the catalyst or decrease in yields.

The following year, Ley and co-workers described a study of the catalytic oxidation of alcohols that used tetraalkylammonium and imidazolium ionic liquids to recover and recycle the catalyst tetrapropylammonium ruthenate(VII).[61] They used *N*-methylmorpholine-*N'*-oxide as an oxidant to oxidize various alcohols in mixtures of CH$_2$Cl$_2$ with either [C$_2$mim][PF$_6$] or [(C$_2$H$_5$)$_4$N]Br. Conversions of over 95% were achieved even after reusing the catalyst/ionic liquid mixture several times.

In 2002, Seddon and Stark reported the palladium-catalysed selective oxidation of benzyl alcohol to benzaldehyde in [C$_4$mim][BF$_4$].[62] The two chemists achieved better reaction rates in the ionic liquid using oxygen as the oxidant compared with rates in the solvent dimethyl sulfoxide. They found, however, that chloride ion impurities in the ionic liquid resulted in the formation of dibenzyl ether as a side product. The selectivity for benzaldehyde was also influenced by the presence of water in the ionic liquid. As water content increased, the yield of benzaldehyde decreased whereas that of benzoic acid increased.

Hydrocarbons can also be oxidized in ionic liquids. In 2006, for example, Tang and co-workers reported the direct conversion of methane into methanol using inorganic platinum catalysts, such as PtCl$_2$ and PtO$_2$, dissolved in a mixture of concentrated sulfuric acid and an ionic liquid.[63] The researchers tested various imidazolium, pyridinium and other types of ionic liquids with Cl$^-$ or [HSO$_4$]$^-$ anions. Ionic liquids with cations containing either no alkyl groups or one or two methyl groups proved effective as media for the catalytic oxidation. The team observed that the ionic liquids not only dissolved the otherwise insoluble platinum catalysts but also played a significant role in promoting the reactivity of the catalysts.

However, ionic liquids with cations containing longer alkyl chains proved not to be suitable media for the conversion of methane as the cations themselves were oxidized in the platinum/acid catalytic system.

10.20 PEPTIDE SYNTHESIS

In 2004, Plaquevent and co-workers reported the first chemical synthesis of peptides in ionic liquids.[64] The group used two triazole coupling agents, **10.71** and **10.72**, with structural similarities to the solvent, the ionic liquid [C$_4$mim][PF$_6$], to couple amino acids such as glycine and phenylalanine. The couplings generated dipeptides and tetrapeptides in good yields and with high purity.

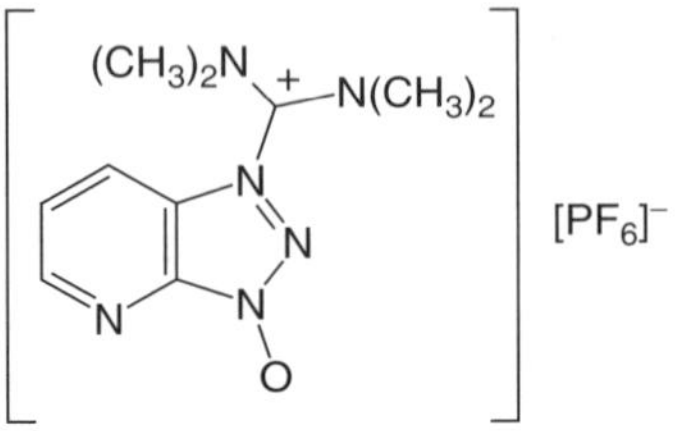

10.71

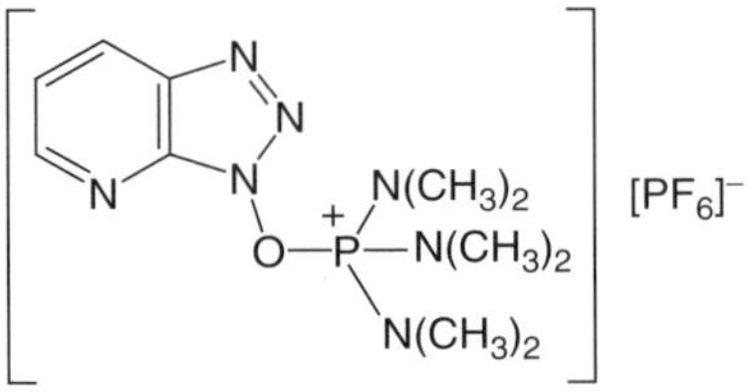

10.72

The following year, Miao and Chan described the use of an ionic liquid functionalized with a hydroxyl group as a soluble support for the synthesis of a oligopeptides.[65] The ionic liquid 3-(2-hydroxyethyl)-1-methylimidazolium tetrafluoroborate (**10.73**) was prepared by the reaction of 1-methylimidazole and 2-bromoethanol. The two chemists used the support for the synthesis of the pentapeptide Leu[5]-enkephalin, H-Tyr-Gly-Gly-Phe-Leu-OH, in "good yield and reasonable purity". Leu[5]-enkephalin is one of two forms of the opioid pentapeptide enkephalin that occurs in the brain to regulate pain in the body.

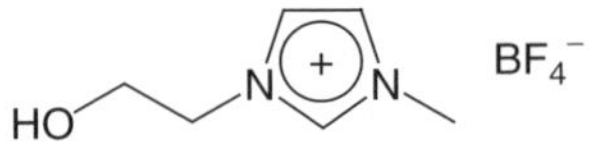

10.73

10.21 PHOTOCHEMICAL REACTIONS

The study of photochemical reactions in ionic liquids has a relatively long history compared with other types of reactions. In general, the polar nature of ionic liquids facilitates their participation in photochemical processes. Imidazolium and pyridinium cations, for example, can accept electrons from photoexcited molecules. The process, which is known as photoinduced electron transfer, furnishes radical ions that react to form the photochemical products.

The earliest report, which appeared in 1978, described the photochemical behaviour of iron(II) complexes with diimine ligands.[66] Diimines are compounds with two carbon–nitrogen double bonds, C=N. The report, by Osteryoung and colleagues, described the use of low-intensity visible light to irradiate iron(II) complexes, with ligands such as $H_3CN=C(CH_3)CH=NCH_3$, in the Lewis acidic ionic liquid $[C_2py]Br$-$AlCl_3$. The iron(II) complexes were converted into iron(III) complexes by the transfer of one electron to the ethylpyridinium cation, $[C_2py]^+$, which reduced to form the dimer **10.74**. The cation therefore participated in the photochemical reaction.

$$H_5C_2N \bigcirc - \bigcirc NC_2H_5$$

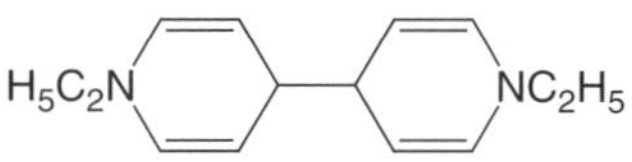

10.74

In 1993, Pagni, Mamantov and co-workers published an investigation into the photochemical behaviour of anthracene in room-temperature chloroaluminate ionic liquids.[67] The photolysis of anthracene (**10.75**) in basic $[C_2mim]Cl$-$AlCl_3$ caused a $[4+4]$ cycloaddition reaction that yielded dimer **10.76**. The same photochemical reaction occurs in conventional organic solvents.

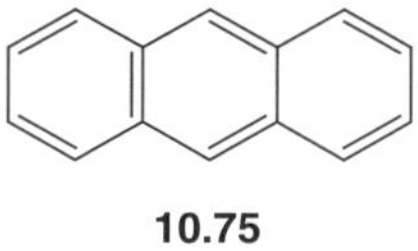

10.75

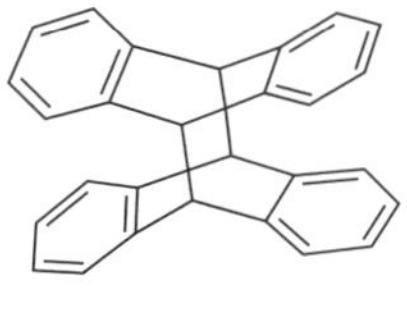

10.76

The cycloaddition only occurred in the basic ionic liquid when it was deoxygenated. Photolysis of anthracene in the oxygenated basic ionic liquid afforded anthraquinone and chlorinated anthracenes.[68]

In acidic $[C_2mim]Cl$-$AlCl_3$ containing the Brønsted acid HCl, photolysis of anthracene resulted in a cascade of oxidation and reduction reactions. The Brønsted acid protonates anthracene to form the anthracenium ion

10.77. On photolysis, anthracene loses an electron to become a radical cation. The electron is transferred to the anthracenium ion, which then readily couples with the radical cation to form various oxidized, reduced and neutral dimeric products. The polar nature of the ionic liquid facilitates the electron transfer.

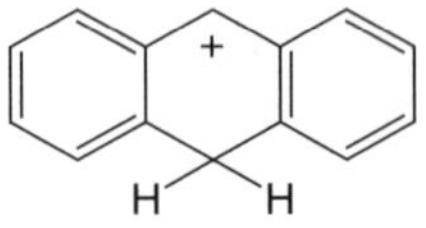

10.77

The choice of ionic liquid can have a pronounced impact on the course of a photochemical reaction. In 2002, for example, Jones and co-workers described an investigation into the use of ammonium-based and imidazolium-based ionic liquids as solvents for the photoreduction of benzophenones by primary amines.[69]

The photoreduction of benzophenone (**10.78**) with *sec*-butylamine, $CH_3CH_2CH(NH_2)CH_3$, in [C$_4$mim][BF$_4$] yielded benzhydrol (**10.79**). In the room-temperature ionic liquids *sec*-butylammonium trifluoroacetate and isopropylammonium nitrate, the photoreduction produced the vicinal diol benzpinacol (**10.80**).

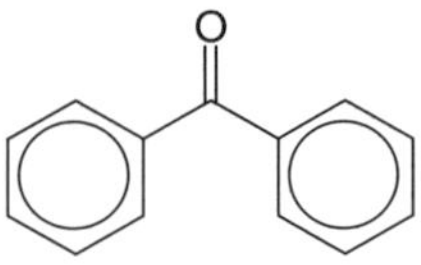

10.78

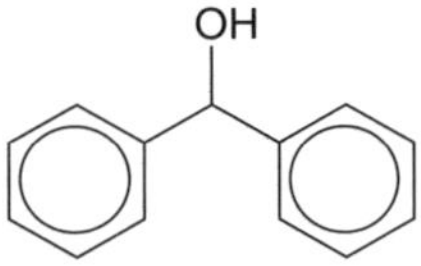

10.79

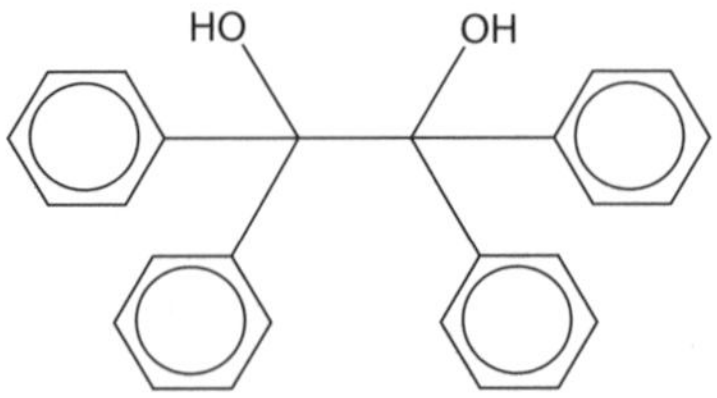

10.80

10.22 POLYMERIZATIONS

A wide variety of polymerizations have been carried out in both chloroaluminate and non-chloroaluminate ionic liquids. Investigations date back to the late 1980s. In 1989, for example, Trivedi demonstrated that benzene and other aromatic hydrocarbons can be electrochemically polymerized in chloroaluminate ionic liquids to give highly conducting films of condensed-ring polymers at the anode.[70]

A few papers on the topic appeared in the 1990s. For instance, in 1992 Kobryanskii and Arnautov reported the use of [C$_4$py]Cl-AlCl$_3$ as a solvent for the oxidative polymerization of benzene to polyphenylenes.[71] They showed that the relative molecular masses of the polymers were higher than those achieved in conventional organic solvents. The reason, they suggested, was the higher solubility of the polymers in ionic liquids, which facilitated the polymerization.

However, as with so many aspects of ionic liquids chemistry, the study of polymerizations in ionic liquids did not blossom until the early 2000s. In 2000, Haddleton and co-workers used [C$_4$mim][PF$_6$] as a solvent for the copper(I)-catalysed living free radical polymerization of methyl methacrylate (**10.81**) to poly(methyl methacrylate) (**10.82**) (Scheme 10.30).[72] The work was the first example of living radical polymerization in an ionic liquid.

Living polymerizations are polymerizations with no termination reactions or in which termination is slow relative to chain growth. They permit the synthesis of highly-branched functionalized macromolecules with accurately controlled molecular architectures and narrow polydispersities, *i.e.* narrow molecular weight distributions. In living polymerizations, reactive polymer intermediates, which can be free radicals, anions or cations, are generated reversibly. These intermediates are either active, when a monomer is added, or dormant until more monomer is added. Because living polymerizations are controlled by addition of a monomer, they are sometimes known as controlled polymerizations.

Methyl methacrylate is soluble in [C$_4$mim][PF$_6$] and can therefore be polymerized in the ionic liquid under homogeneous conditions.

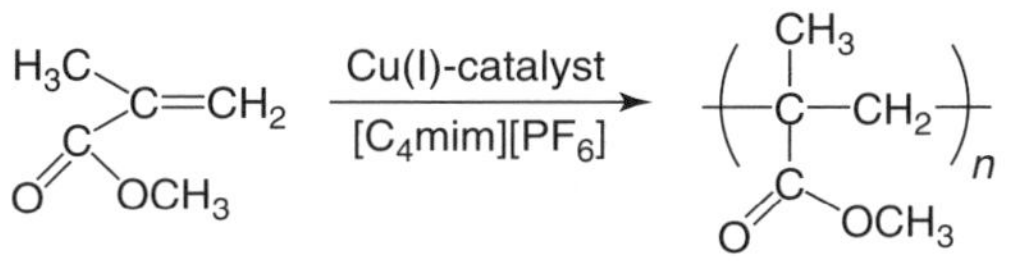

Scheme 10.30   Polymerization of methyl methacrylate.

Haddleton's team showed not only that rate of living radical polymerization of methyl methacrylate in the ionic liquid was fast but also that the products could easily be extracted from the ionic liquid with toluene. Furthermore, the products exhibited narrow polydispersity and were not contaminated with the transition-metal catalyst.

In 2002, Brazel, Rogers and co-workers reported that the monomers methyl methacrylate and styrene undergo free radical polymerization in [C$_4$mim][PF$_6$] using conventional organic initiators such as AIBN (2,2'-azobisisobutyronitrile), (CH$_3$)$_2$C(CN)N=NC(CH$_3$)$_2$CN, or benzoyl peroxide, (C$_6$H$_5$CO)$_2$O$_2$.[73] The reactions, which were rapid, were terminated by quenching into methanol or aqueous ethanol. The monomer and initiators were soluble in the ionic liquid. The polymer, however, was not and phase-separated soon after the polymerization started. It was then isolated by filtration. The authors suggested that the process is attractive for "green" polymer synthesis.

A team led by Watanabe showed that polymer electrolytes with high ionic conductivities at room temperature can be prepared by the radical polymerization of vinyl monomers, such as methyl methacrylate, in the ionic liquid [C$_2$mim][NTf$_2$], using benzoyl peroxide as an initiator.[74] The products, which the group dubbed "ion gels", had sufficient mechanical strength, transparency and flexibility for use in a wide variety of solid-state electrochemical devices.

Monomers such as butyl acrylate are insoluble in [C$_4$mim][PF$_6$]. Biedroń and Kubisa demonstrated, in 2001, that the ionic liquid could be used for biphasic copper(I)-catalysed living radical polymerizations of such monomers.[75] In this system, the monomer formed a separate phase from the catalyst-containing ionic liquid.

The following year, Shaughnessy, Rogers and co-workers reported that mixtures of methanol and polar, non-coordinating ionic liquids such as [C$_6$py][NTf$_2$] are effective solvents for the palladium-catalysed copolymerization of styrene and CO (Scheme 10.31).[76] The

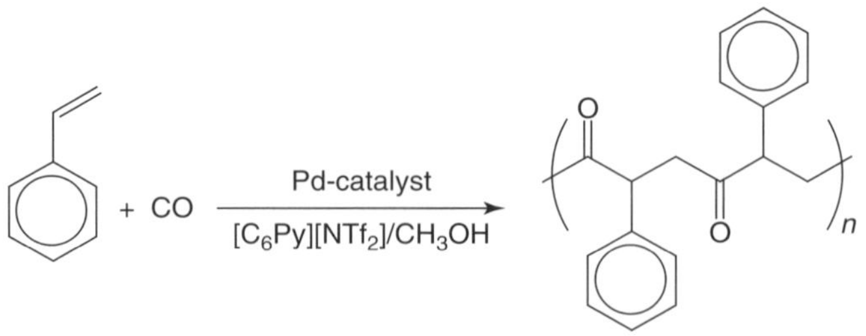

Scheme 10.31 Copolymerization of styrene and carbon monoxide.

copolymerizations, which were carried out at 70 °C, resulted in improved yields and increased molecular weights compared with the use of methanol as solvent for the polymerizations.

In the same year, Nobile and colleagues described the rhodium(I)-catalysed polymerization of phenylacetylene, C_6H_5-C≡CH, in the presence of the co-catalyst triethylamine using the ionic liquids [C_4py][BF_4] or [C_4mim][BF_4] as solvent.[77] The rhodium complex in both ionic liquids was recycled without significant loss of activity.

In 2004, 15 years after the early work by Trivedi[70] (see above) using chloroaluminate ionic liquids, Endres and colleagues described the use of Lewis neutral ionic liquids for the electropolymerization of benzene.[78] The team used [C_6mim][FAP] and [1-butyl-1-methylpyrrolidinium][NTf_2] as electrolytes for the oxidation of benzene on a platinum anode. A black film of poly(para)phenylene was deposited on the electrode.

Ring-opening polymerizations can also be carried out in ionic liquids. In one study, Schubert and co-workers investigated the living cationic ring-opening polymerization of 2-ethyl-2-oxazoline (**10.83**) in the hydrophobic ionic liquid [C_4mim][PF_6] under microwave irradiation (Scheme 10.32).[79] The rate of polymerization proved to be higher than in common organic solvents such as acetonitrile and ethyl ether. In addition, the polymer was efficiently extracted with water, allowing the hydrophobic ionic liquid to be recovered and used again. The process was therefore "green" to the extent that it avoided the use of volatile organic compounds, the authors noted.

Not all attempts to polymerize monomers in ionic liquids have proved successful. For example, in some of the earliest work in this field, published in 1990, Carlin and co-workers studied the Ziegler–Natta polymerization of ethylene in acidic [C_2mim]Cl-$AlCl_3$.[80] Ziegler–Natta catalysts are used to polymerize alk-1-enes. They typically consist of a mixture of a titanium compound and an organoaluminium compound. Carlin's team employed a combination of $TiCl_4$ and $(CH_3)_3Al_2Cl_3$ but achieved only low yields of polyethylene using the ionic liquid.

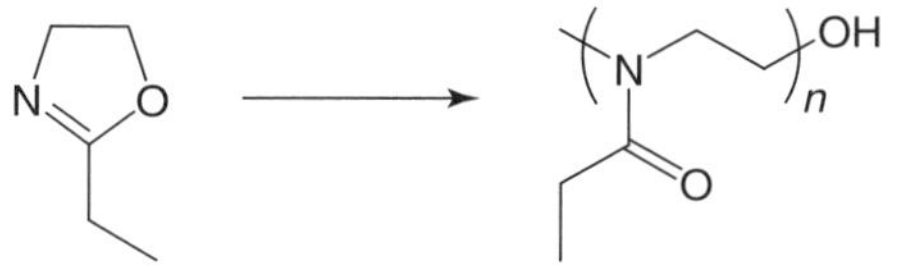

10.83

Scheme 10.32 Ring-opening polymerization.

In another study, Shaughnessy, Rogers and co-workers showed that the activity of a cationic palladium catalyst for the polymerization of ethylene is an order of magnitude lower in the ionic liquids [C$_4$mim][PF$_6$] and [C$_6$pyr][NTf$_2$] than in dichloromethane.[81]

Ionic liquids can function not only as solvents for polymerizations but also as additives. In 2002, for example, Brazel, Rogers and co-workers reported that [C$_4$mim][PF$_6$] is an efficient plasticiser for PMMA, poly (methyl methacrylate).[82] Plastics are solid materials, typically synthetic polymers, that become mobile when heated and can therefore be cast into moulds, extruded to form rods or tubes, or used to form laminated products and surface coatings. Plasticisers are added to hard synthetic polymers to make them softer, more flexible, more workable and more durable. The 2002 investigation showed that the physical characteristics of PMMA plasticized with [C$_4$mim][PF$_6$] are comparable with those of PMMA containing conventional plasticizers such as dioctyl phthalate.

Ionic liquids can also act as surfactants to stabilize polymer beads synthesized by suspension polymerization. Suspension polymerization typically employs an aqueous suspension of droplets of an organic phase containing the monomers and polymerization initiator. Schubert's group tested three water-soluble imidazolium ionic liquids, [C$_4$mim]Cl, [C$_{10}$mim]Cl and [C$_{16}$mim]Cl, as potential stabilizers for crosslinked polymer beads prepared by the copolymerization of ethylene glycol methacrylate (**10.84**) and *N*-vinylimidazole (**10.85**) in toluene using AIBN as the initiator (Scheme 10.33).[83]

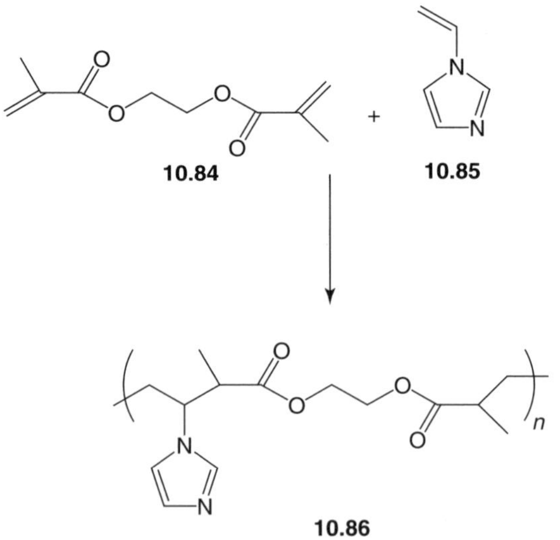

Scheme 10.33   Synthesis of a crosslinked polymer.

The resulting crosslinked polymer, poly(ethylene glycol dimethacrylate-*N*-vinylimidazole) (**10.86**), was insoluble in the aqueous phase which contained the ionic liquid. The average particle size of the polymer beads ranged from the nano- to the macroscale depending on the concentration and the length of the aliphatic side-chain of the ionic liquid cations. As the concentration of a specific ionic liquid increased, the average particle size decreased. Similarly, as the alkyl chain length increased, the average particle size decreased.

REFERENCES

1. C. P. Mehnert, N. C. Dispenziere and R. A. Cook, *Chem. Commun.*, 2002, 1610.
2. S. Abelló, F. Medina, X. Rodríguez, Y. Cesteros, P. Salagre, J. E. Sueiras, D. Tichit and B. Coq, *Chem. Commun.*, 2004, 1096.
3. M. J. Earle, P. B. McCormac and K. R. Seddon, *Chem. Commun.*, 1998, 2245.
4. W. Chen, L. Xu, C. Chatterton and J. Xiao, *Chem. Commun.*, 1999, 1247.
5. G. W. Kabalka, G. Dong and B. Venkataiah, *Org. Lett.*, 2003, **5**, 893.
6. C. J. Adams, M. J. Earle and K. R. Seddon, *Green Chem.*, 2000, **2**, 21.
7. A. Kamimura and S. Yamamoto, *Org. Lett.*, 2007, **9**, 2533.
8. Y. Wang, H. Li, C. Wang and H. Jiang, *Chem. Commun.*, 2004, 1938.
9. J. Peng and Y. Deng, *New J. Chem.*, 2001, **25**, 639.
10. H. Yang, Y. Gu, Y. Deng and F. Shi, *Chem. Commun.*, 2002, 274.
11. H. Kawanami, A. Sasaki, K. Matsui and Y. Ikushima, *Chem. Commun.*, 2003, 896.
12. R. J. C. Brown, P. J. Dyson, D. J. Ellis and T. Welton, *Chem. Commun.*, 2001, 1862.
13. Y. Chauvin, B. Gilbert and I. Guibard, *J. Chem. Soc., Chem. Commun.*, 1990, 1715.
14. B. Ellis, W. Keim and P. Wasserscheid, *Chem. Commun.*, 1999, 337.
15. P. Wasserscheid, C. M. Gordon, C. Hilgers, M. J. Muldoon and I. R. Dunkin, *Chem. Commun.*, 2001, 1186.
16. J. E. L. Dullius, P. A. Z. Suarez, S. Einloft, R. F. de Souza and J. Dupont, *Organometallics*, 1998, **17**, 815.

17. R. A. Ligabue, J. Dupont and R. F. de Souza, *J. Mol. Catal. A: Chem.*, 2001, **169**, 11.
18. J. Zimmermann, P. Wasserscheid, I. Tkatchenko and S. Stutzmann, *Chem. Commun.*, 2002, 760.
19. C. E. Song and E. J. Roh, *Chem. Commun.*, 2000, 837.
20. M. M. Abu-Omar, G. S. Owens and A. Durazo, in *Ionic Liquids as Green Solvents: Progress and Prospects*, ed. R. D. Rogers and K. R. Seddon, ACS Symposium Series 856, American Chemical Society, Washington DC, 2003, p. 277.
21. K.-H. Tong, K.-Y. Wong and T. H. Chan, *Org. Lett.*, 2003, **5**, 3423.
22. R. Bernini, E. Mincione, A. Coratti, G. Fabrizi and G. Battistuzzi, *Tetrahedron*, 2004, **60**, 967.
23. H.-P. Zhu, F. Yang, J. Tang and M.-Y. He, *Green Chem.*, 2003, **5**, 38.
24. C. Imrie, E. R. T. Elago, C. W. McCleland and N. Williams, *Green Chem.*, 2002, **4**, 159.
25. A. R. Gholap, K. Venkatesan, T. Daniel, R. J. Lahoti and K. V. Srinivasan, *Green Chem.*, 2003, **5**, 693.
26. T. P. Wells, J. P. Hallett, C. K. Williams and T. Welton, *J. Org. Chem.*, 2008, **73**, 5585.
27. F. Shi, H. Xiong, Y. Gu, S. Guo and Y. Deng, *Chem. Commun.*, 2003, 1054.
28. C. Chiappe, D. Capraro, V. Conte and D. Pieraccini, *Org. Lett.*, 2001, **3**, 1061.
29. R. Rajagopal, D. V. Jarikote, R. J. Lahoti, T. Daniel and K. V. Srinivasan, *Tetrahedron Lett.*, 2003, **44**, 1815.
30. M. J. Earle, S. P. Katdare and K. R. Seddon, *Org. Lett.*, 2004, **6**, 707.
31. G. W. Parshall, *J. Am. Chem. Soc.*, 1972, **94**, 8716.
32. J. F. Knifton (to Texaco Inc.), *U. S. Pat.* 4451679 (May 29, 1984).
33. Y. Chauvin, L. Mussmann and H. Olivier, *Angew. Chem. Int. Ed. Engl.*, 1995, **34**, 2698.
34. C. P. Mehnert, R. A. Cook, N. C. Dispenziere and M. Afeworki, *J. Am. Chem. Soc.*, 2002, **124**, 12932.
35. L. Leclercq, I. Suisse and F. Agbossou-Niedercorn, *Chem. Commun.*, 2008, 311.
36. P. A. Z. Suarez, J. E. L. Dullius, S. Einloft, R. F. de Souza and J. Dupont, *Polyhedron*, 1996, **15**, 1217.
37. A. L. Monteiro, F. K. Zinn, R. F. de Souza and J. Dupont, *Tetrahedron: Asymmetry*, 1997, **8**, 177.
38. P. J. Dyson, D. J. Ellis, D. G. Parker and T. Welton, *Chem. Commun.*, 1999, 25.

39. H.-T. Wong, Y. H. See-Toh, F. C. Ferreira, R. Crook and A. G. Livingston, *Chem. Commun.*, 2006, 2063.
40. J. Dupont, G. S. Fonseca, A. P. Umpierre, P. F. P. Fichtner and S. R. Teixeira, *J. Am. Chem. Soc.*, 2002, **123**, 4228.
41. E. T. Silveira, A. P. Umpierre, L. M. Rossi, G. Machado, J. Morais, G. V. Soares, I. J. R. Baumvol, S. R. Teixeira, P. F. P. Fichtner and J. Dupont, *Chem. Eur. J.*, 2004, **10**, 3734.
42. R. A. Brown, P. Pollet, E. McKoon, C. A. Eckert, C. L. Liotta and P. G. Jessop, *J. Am. Chem. Soc.*, 2001, **123**, 1254.
43. F. Liu, M. B. Abrams, R. T. Baker and W. Tumas, *Chem. Commun.*, 2001, 433.
44. J. Arras, M. Steffan, Y. Shayeghi and P. Claus, *Chem. Commun.*, 2008, 4058.
45. T. J. Geldbach, D. Zhao, N. C. Castillo, G. Laurenczy, B. Weyershausen and P. J. Dyson, *J. Am. Chem. Soc.*, 2006, **128**, 9773.
46. K. K. Laali and V. J. Gettwert, *J. Org. Chem.*, 2001, **66**, 35.
47. S. T. Handy and C. R. Egrie, in *Ionic Liquids: Industrial Applications to Green Chemistry*, ed. R. D. Rogers and K. R. Seddon, ACS Symposium Series 818, American Chemical Society, Washington DC, 2002, p. 134..
48. N. L. Lancaster and V. Llopis-Mestre, *Chem. Commun.*, 2003, 2812.
49. C. Wheeler, K. N. West, C. L. Liotta and C. A. Eckert, *Chem. Commun.*, 2001, 887.
50. J. X. Wu, B. Beck and R. X. Ren, *Tetrahedron Lett.*, 2002, **43**, 387.
51. W.-C. Shieh, M. Lozanov and O. Repič, *Tetrahedron Lett.*, 2003, **44**, 6943.
52. N. M. T. Lourenço and C. A. M. Afonso, *Tetrahedron*, 2003, **59**, 789.
53. D. B. G. Williams, M. Ajam and A. Ranwell, *Organometallics*, 2006, **25**, 3088.
54. R. C. Buijsman, E. van Vuuren and J. G. Sterrenburg, *Org. Lett.*, 2001, **3**, 3785.
55. N. Audic, H. Clavier, M. Mauduit and J.-C. Guillemin, *J. Am. Chem. Soc.*, 2003, **125**, 9248.
56. A. L. Miller II and N. B. Bowden, *Chem. Commun.*, 2007, 2051.
57. S. Csihony, C. Fischmeister, C. Bruneau, I. T. Horváth and P. H. Dixneuf, *New. J. Chem.*, 2002, **26**, 1667.
58. C. Thurier, C. Fischmeister, C. Bruneau, H. Olivier-Bourbigou and P. H. Dixneuf, *ChemSusChem*, 2008, **1**, 118.
59. J. Muzart, *Adv. Synth. Catal.*, 2006, **348**, 275.

60. J. Howarth, *Tetrahedron Lett.*, 2000, **41**, 6627.
61. S. V. Ley, C. Ramarao and M. D. Smith, *Chem. Commun.*, 2001, 2278.
62. K. R. Seddon and A. Stark, *Green Chem.*, 2002, **4**, 119.
63. J. Cheng, Z. Li, M. Haught and Y. Tang, *Chem. Commun.*, 2006, 4617.
64. H. Vallette, L. Ferron, G. Coquerel, A.-C. Gaumont and J.-C. Plaquevent, *Tetrahedron Lett.*, 2004, **45**, 1617.
65. W. Miao and T.-H. Chan, *J. Org. Chem.*, 2005, **70**, 3251.
66. H. L. Chum, D. Koran and R. A. Osteryoung, *J. Am. Chem. Soc.*, 1978, **100**, 310.
67. G. Hondrogiannis, C. W. Lee, R. M. Pagni and G. Mamantov, *J. Am. Chem. Soc.*, 1993, **115**, 9828.
68. R. M. Pagni, in *Ionic Liquids as Green Solvents: Progress and Prospects*, ed. R. D. Rogers and K. R. Seddon, ACS Symposium Series 856, American Chemical Society, Washington DC, 2003, p. 344.
69. J. L. Reynolds, K. R. Erdner and P. B. Jones, *Org. Lett.*, 2002, **4**, 917.
70. D. C. Trivedi, *J. Chem. Soc., Chem. Commun.*, 1989, 544.
71. V. M. Kobryanskii and S. A. Arnautov, *J. Chem. Soc., Chem. Commun.*, 1992, 727.
72. A. J. Carmichael, D. M. Haddleton, S. A. F. Bon and K. R. Seddon, *Chem. Commun.*, 2000, 1237.
73. K. Hong, H. Zhang, J. W. Mays, A. E. Visser, C. S. Brazel, J. D. Holbrey, W. M. Reichert and R. D. Rogers, *Chem. Commun.*, 2002, 1368.
74. M. A. B. H. Susan, T. Kaneko, A. Noda and M. Watanabe, *J. Am. Chem. Soc.*, 2005, **127**, 4976.
75. T. Biedroń and P. Kubisa, *Macromol, Rapid Commun.*, 2001, **22**, 1237.
76. M. A. Klingshirn, G. A. Broker, J. D. Holbrey, K. H. Shaughnessy and R. D. Rogers, *Chem. Commun.*, 2002, 1394.
77. P. Mastrorilli, C. F. Nobile, V. Gallo, G. P. Suranna and G. Farinola, *J. Mol. Catal. A: Chem.*, 2002, **184**, 73.
78. S. Z. El Abedin, N. Borissenko and F. Endres, *Electrochem. Commun.*, 2004, **6**, 422.
79. C. Guerrero-Sanchez, R. Hoogenboom and U. S. Schubert, *Chem. Commun.*, 2006, 3797.
80. R. T. Carlin, R. A. Osteryoung, J. S. Wilkes and J. Rovang, *Inorg. Chem.*, 1990, **29**, 3003.
81. K. H. Shaughnessy, M. A. Klingshirn, S. J. P'Pool, J. D. Holbrey and R. D. Rogers, in *Ionic Liquids as Green Solvents Progress and Prospects*, ed. R. D. Rogers and K. R. Seddon, ACS

Symposium Series 856, American Chemical Society, Washington DC, 2003, p. 300.

82. M. P. Scott, C. S. Brazel, M. G. Benton, J. W. Mays, J. D. Holbrey and R. D. Rogers, *Chem. Commun.*, 2002, 1370.
83. C. Guerrero-Sanchez, T. Erdmenger, P. Šereda, D. Wouters and U. S. Schubert, *Chem. Eur. J.*, 2006, **12**, 9036.

CHAPTER 11

Named Organic Reactions

11.1 BAEYER–VILLIGER OXIDATION

The Baeyer–Villiger oxidation is the oxidation of acyclic ketones to esters or cyclic ketones to lactones. The reactions are typically carried out in chlorinated solvents, such as dichloromethane, using hydrogen peroxide or a peroxyacid as the oxidant.

In 2003, Bernini and co-workers reported the Baeyer–Villiger oxidation of ten cyclic ketones in [C$_4$mim][BF$_4$].[1] They carried out the reactions using an aqueous solution of hydrogen peroxide as the oxidant and methyltrioxorhenium, CH$_3$ReO$_3$, as a recyclable catalyst. The conversions of the ketones and the yields of the lactones were generally variable although in some cases high. For example, the oxidation of cyclobutanone (**11.1**) resulted in a quantitative yield of γ-butyrolactone (**11.2**) (Scheme 11.1).

11.2 BAYLIS–HILLMAN REACTION

In the Baylis–Hillman reaction, a carbon–carbon single bond is formed between the α-position of the C=C double bond in a conjugated

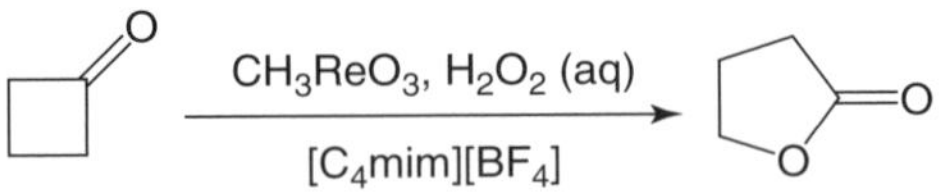

11.1   CH$_3$ReO$_3$, H$_2$O$_2$ (aq)   **11.2**

[C$_4$mim][BF$_4$]

Scheme 11.1   Baeyer–Villiger oxidation.

An Introduction to Ionic Liquids
By Michael Freemantle
© Michael Freemantle 2010
Published by the Royal Society of Chemistry, www.rsc.org

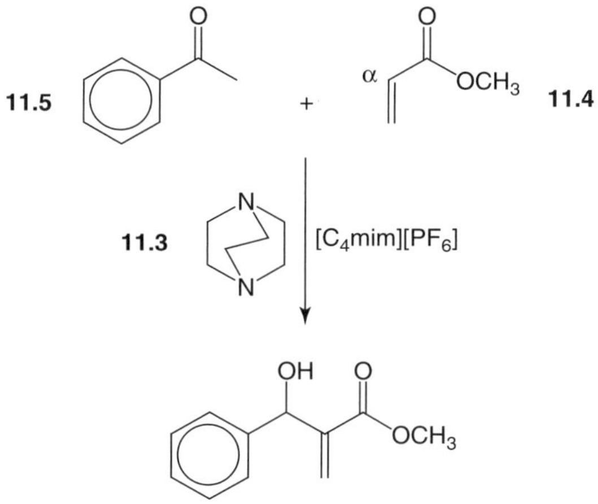

Scheme 11.2 Baylis–Hillman reaction.

carbonyl compound, *e.g.* an amide or an ester, and a carbon electrophile, such as an aldehyde, using a nucleophilic catalyst. The products are allylic alcohols. The catalysts are typically tertiary amines such as 1,4-diazabicyclo[2.2.2]octane (**11.3**), more commonly known as DABCO.

In 2001, Rosa and co-workers described investigations of various DABCO-catalysed Baylis–Hillman reactions in [C$_4$mim][BF$_4$] and [C$_4$mim][PF$_6$].[2] The reaction of methyl acrylate (**11.4**) and benzaldehyde (**11.5**) to yield the alkoxyester **11.6**, for example (Scheme 11.2), was found to be more than 30 times faster in [C$_4$mim][PF$_6$] than in acetonitrile.

The first example of the use of chiral ionic liquids as reaction media for the asymmetric Baylis–Hillman reaction was reported by Vo-Thanh and co-workers in 2004.[3] The team carried out the DABCO-catalysed Baylis–Hillman reaction of methyl acrylate and benzaldehyde using the chiral ionic liquid **11.7**, which has a triflate anion and a cation derived from (–)-*N*-methylephedrine. Yields of the alkoxyester of up to 88% were achieved with enantiomeric excesses of up to 44%.

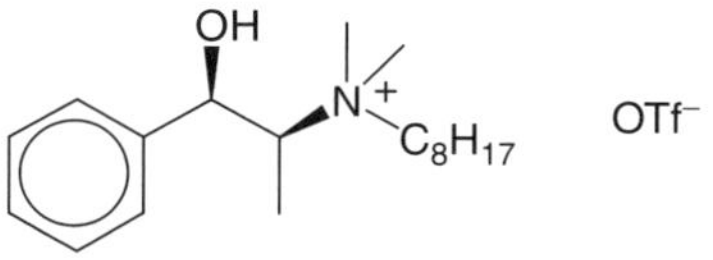

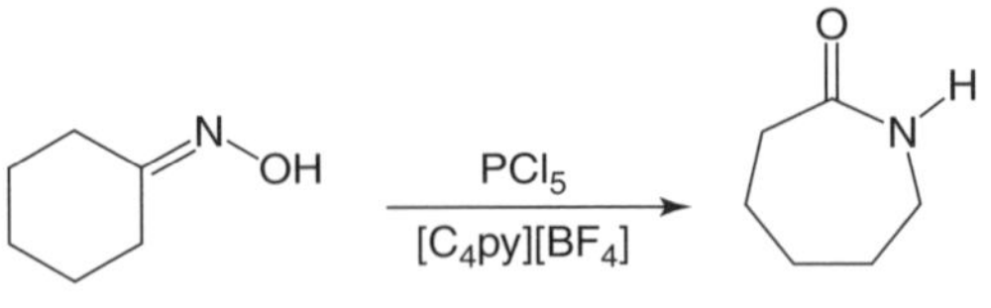

11.9   **11.8**

Scheme 11.3   Beckmann rearrangement.

11.3   BECKMANN REARRANGEMENT

The Beckmann rearrangement is an acid-catalysed process that converts ketoximes and aldoximes into their corresponding amides. Oximes have the general formula $R_1R_2C=N$-OH, where R_1 is organic while R_2 is organic for ketoximes and hydrogen for aldoximes. Cyclic ketoximes are converted into lactams by the process. The rearrangement is used for the manufacture of ε-caprolactam **11.8**, a monomer that is polymerized in the production of nylon 6.

In 2001, Peng and Deng reported an investigation into the Beckmann rearrangements of three ketoximes in catalyst-containing ionic liquids.[4] They tested PCl_5 and other phosphorus compounds as catalysts in the ionic liquids $[C_4mim][BF_4]$, $[C_4mim][CF_3CO_2]$ and $[C_4py][BF_4]$. The reactions were carried out without the addition of organic solvents. They observed "excellent" conversion of cyclohexanone oxime (**11.9**, and selectivity for 11.8 (Scheme 11.3)). For example, PCl_5 and $[C_4py][BF_4]$ resulted in over 97% conversion of the oxime, with over 96% selectivity for the lactam. The turnover number was 6.5.

11.4   DIELS–ALDER REACTION

The Diels–Alder reaction is a [4 + 2]-cycloaddition reaction in which a conjugated diene reacts with a compound containing a carbon–carbon double bond (C=C), known as a dienophile. The diene, which contains the sequence -C=C-C=C-, undergoes 1,4-cycloaddition (addition at its C-1 and C-4 positions) to form a product, known as an adduct, containing a cyclohexene ring. The reaction is widely used in organic chemistry and in the pharmaceutical, plastics, textiles and other industries to synthesize various intermediates and products.

The first report of the use of an ionic liquid for a Diels–Alder reaction was published by Jaeger and Trucker in 1989.[5] They showed that the reaction of cyclopentadiene (**11.10**) with the dienophile methyl acrylate, $CH_2=CHCO_2CH_3$, in the neutral ionic liquid ethylammonium nitrate $[C_2H_5NH_3][NO_3]$ produced a mixture of bridged *exo*- and *endo*- adducts:

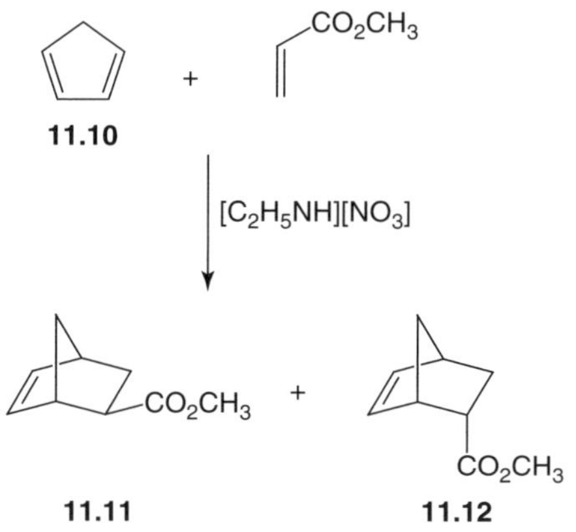

Scheme 11.4 Diels–Alder reaction of cyclopentadiene and methyl acrylate.

11.11 and **11.12**, respectively (Scheme 11.4). The prefix *exo-* is used to describe products in which the dienophile substituents lie on the same side as the bridge. The substituents of *endo*-products lie on the other side of the cyclohexene ring.

In general, Diels–Alder reactions yield *endo*-products although it is well established that selectivity for the *endo-* or *exo*-products is strongly influenced by the solvent. Jaeger and Trucker reported that the reaction of cyclopentadiene and methyl acrylate in [$C_2H_5NH_3$][NO_3] was around three times faster than in benzene and that the ratio of *endo-* to *exo*-products was 6.7 : 1 compared with 2.8 : 1 for benzene. The reactivity and *endo*-selectivity, however, were lower than those for water.

Nearly a decade later, Howarth and co-workers showed that dialkylimidazolium ionic liquids could act as weak Lewis acids for the catalytic Diels–Alder reaction of cyclopentadiene with crotonaldehyde, $CH_3CH=CHCHO$, or methacrolein, $CH_2=C(CH_3)CHO$.[6] The *endo:exo* selectivities for the products of the reaction with crotonaldehyde exceeded 90 : 10 whereas the reaction with methacrolein favoured the *exo*-adduct with selectivities in excess of 15 : 85.

The investigations of Diels–Alder reactions in ionic liquids by Howarth's team included one of the earliest attempts to use an ionic liquid (**11.13**) with a chiral cation for asymmetric synthesis. However, the attempts proved unsuccessful with enantiomeric excesses of less than 5%.

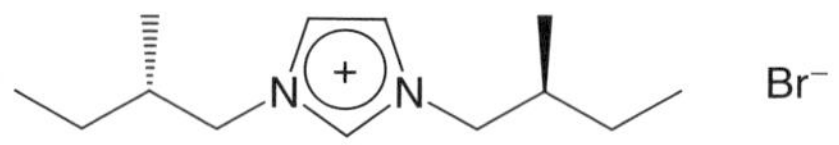

11.13

In 1999, Lee reported that the *endo*-selectivity and rate of the Diels–Alder reaction of cyclopentadiene and methyl acrylate could be significantly enhanced by using acidic [C$_2$mim]Cl-AlCl$_3$ or acidic [C$_4$py]Cl-AlCl$_3$ as a solvent and catalyst.[7] The rates of reaction in the chloroaluminate ionic liquids were more than double and the *endo*-selectivity almost ten times better than in water.

The same year, Earle and co-workers showed that neutral ionic liquids, other than [C$_2$H$_5$NH$_3$][NO$_3$], are "excellent" solvents for Diels–Alder reactions.[8] For example, the reaction of cyclopentadiene and ethyl acrylate, CH$_2$=CHCO$_2$C$_2$H$_5$, in [C$_4$mim][BF$_4$] resulted in a 99% yield with an *endo : exo* ratio of 5.0 : 1. They also showed that, for the Diels–Alder reaction of isoprene (**11.14**) and methyl vinyl ketone (**11.15**) in [C$_4$mim][PF$_6$], the addition of a small amount of zinc iodide, a mild Lewis acid, to the reaction mixture improved the selectivity for the 4-isomer product **11.16** over the 2-isomer product **11.17** from 4 : 1 to 20 : 1 (Scheme 11.5).

Earle and colleagues noted that the ionic liquids are safe to use, thermally robust, recyclable and could potentially be used on an industrial scale for Diels–Alder reactions. However, Tiwari and Kumar subsequently showed that the rates of the Diels–Alder reactions of ethyl acrylate and other alkyl acrylates with cyclopentadiene are several times faster in water than in [C$_4$mim][BF$_4$] and other neutral imidazolium ionic liquids.[9] They concluded that room-temperature ionic liquids "are not as effective as water in promoting Diels–Alder reactions."

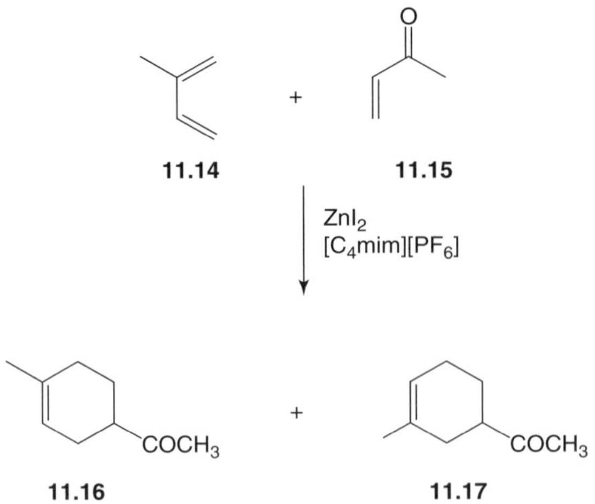

Scheme 11.5   Diels–Alder reaction of isoprene and methyl vinyl ketone.

Scheme 11.6 Sc(OTf)$_2$-catalysed Diels–Alder reaction.

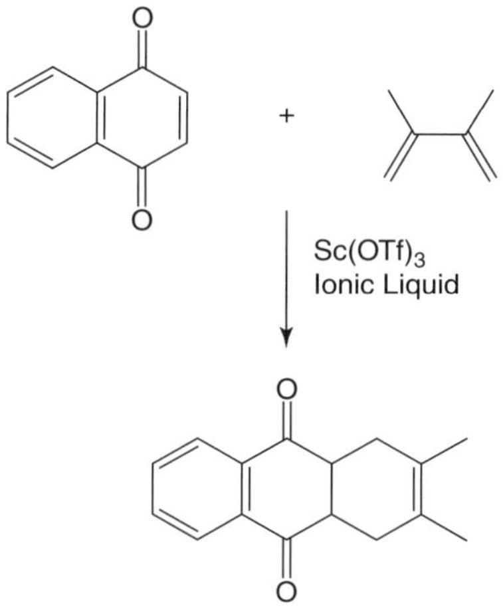

Song, Choi and co-workers reported, in 2001, that imidazolium ionic liquids are "powerful media" for Diels–Alder reactions catalysed by scandium triflate, Sc(OTf)$_3$.[10] Ionic liquids such as [C$_4$mim][PF$_6$] not only accelerated the reaction rates but also improved selectivity for *endo*-adducts where the reaction products are bridged compounds. They also showed that the catalyst-containing ionic liquid can be readily recovered and reused several times without loss of activity. Furthermore yields were high. For example, the Sc(OTf)$_3$-catalysed Diels–Alder reaction of 2,3-dimethylbuta-1,3-diene (**11.18**) with the dienophile 1,4-naphthoquinone (**11.19**) in the ionic liquids [C$_4$mim][PF$_6$], [C$_4$mim]SbF$_6$] or [C$_4$mim][OTf] resulted in product **11.20** in yields of more than 99% (Scheme 11.6).

11.5 FRIEDEL–CRAFTS REACTIONS

Friedel–Crafts reactions are electrophilic substitution reactions that are widely used to introduce an alkyl or acyl group into an aromatic ring. In a typical reaction, a hydrogen atom on an aromatic ring is substituted with an alkyl group, such as methyl or ethyl, or an acyl group, *e.g.* acetyl or benzoyl. Aluminium chloride, a Lewis acid, is commonly used as a catalyst for the reaction. During the reaction, the aromatic ring is oxidized by the removal of a π-electron and an intermediate carbonium ion forms as a result.

The reactions are conventionally carried out using $AlCl_3$ dissolved or suspended in an inert organic molecular solvent. In 1986, Wilkes and co-workers showed that the acidic form of the imidazolium chloro-aluminate room-temperature ionic liquid [C_2mim]Cl-$AlCl_3$ acts as the solvent and catalyst for the Friedel–Crafts alkylation and acylation of benzene.[11]

For the alkylations, alkyl halides such as methyl chloride were added to a mixture of the ionic liquid, containing excess $AlCl_3$ to make it acidic, and benzene under anhydrous conditions. Various mono- and polyalkylated benzenes were produced. Acetylation of benzene in the acidic ionic liquid using acetyl chloride, CH_3COCl, as the acylating agent yielded acetophenone, $C_6H_5COCH_3$.

No substitutions occurred when basic imidazolium chloroaluminate ionic liquids were used. The basic mixtures contained excess [C_2mim]Cl.

The chemists concluded that the Lewis acid $Al_2Cl_7^-$ in the acidic ionic liquids was the catalyst responsible for promoting the reactions. In the acetylations, for example, they suggested that the catalyst reacted with CH_3COCl to form the electrophile CH_3CO^+:

$$CH_3COCl + Al_2Cl_7^- \rightarrow CH_3CO^+ + 2AlCl_4$$

Subsequent studies using infrared spectroscopy revealed that the Friedel–Crafts acetylation of benzene with acetyl chloride in acidic [C_4mim]Cl-$AlCl_3$ is exactly the same as in the molecular organic solvent 1,2-dichloroethane.[12] Acetyl chloride and $AlCl_3$ were shown to combine to form the key intermediate $[CH_3CO]^+[AlCl_4]^-$, which reacted with benzene to form acetophenone.

Alkylations of aromatic compounds using olefins instead of alkyl halides as alkylating agents and acidic chloroaluminate ionic liquids as catalysts for the synthesis of linear alkylbenzenes were patented in 1995.[13] Because of their polar nature, acidity and density, the ionic liquid alkylation catalysts readily formed a separate phase from the other components of the reaction mixture, according to the inventors. The catalysts could therefore be recycled.

The catalyst $AlCl_3$, on the other hand, cannot be recovered and reused although it used in large amounts in industry for catalysing Friedel–Crafts alkylations and acylations. The catalyst is destroyed before the products are recovered and, as a result, large volumes of hydrated aluminium waste are generated. The possibility of recovering and recycling Friedel–Crafts catalysts could therefore prove important in industry, *e.g.* in the synthesis of linear alkylbenzenes, which are produced on a large scale for the manufacture of detergents.

However, studies, reported in 2002, on the performance of chloroaluminate ionic liquids for the alkylation of benzene with ethylene for the manufacture of ethylbenzene revealed that, although the ionic liquids offer the advantages of improved recovery and recyclability, they are inferior in terms of activity, selectivity and cost when compared with commercial catalysts such as $AlCl_3$.[14]

Friedel–Crafts acylations are used in industry to produce aromatic ketones. These compounds are important intermediates for the manufacture of agrochemicals, dyes, fragrances, pharmaceuticals and other chemical products.

In a paper published in 1998, Seddon and co-workers showed that acidic compositions of $[C_2mim]Cl$-$AlCl_3$ are "excellent" reaction media for performing Friedel–Crafts acylations of simple aromatic compounds such as toluene, $C_6H_5CH_3$, anisole, $C_6H_5OCH_3$, and chlorobenzene, C_6H_5Cl.[15] The yields and selectivities were comparable to the best literature yields and selectivities achieved in conventional molecular solvents.

In 2002, Ross and Xiao reported that Friedel–Crafts acetylations and benzoylations of aromatic compounds such as acetophenone can be carried out efficiently in the non-chloroaluminate ionic liquid $[C_4mim][BF_4]$ using the metal triflate $Cu(OTf)_2$ as the Lewis acid catalyst.[16] Earle and co-workers subsequently showed that metal bistriflamide compounds such as $Sn(NTf_2)_2$ are good Friedel–Crafts acylation catalysts either in the absence of any solvent or when an ionic liquid such as $[C_4mim][NTf_2]$ is used as the reaction medium.[17] In the latter case, the catalyst is generated *in situ* by dissolving a metal salt, *e.g.* $SnCl_4$, in $[C_4mim][NTf_2]$. The ionic liquid allows the catalyst to be readily recovered, recycled and reused.

Ionic liquids have also been used for Friedel–Crafts reactions other than alkylations and acylations. For example, in 2004 Wang and Wang carried out the Friedel–Crafts reaction of PCl_3, a liquid at room temperature, with benzene in a chloroaluminate ionic liquid to synthesize dichlorophenylphosphine (Section 9.2).[18]

In another example, Salunkhe and co-workers showed that acidic $[C_4mim]Cl$-$AlCl_3$ ionic liquids can be used as reaction media and Lewis acid catalysts for the Friedel–Crafts sulfonylation reaction of benzene and substituted benzenes like mesitylene (**11.21**) with 4-methylbenzenesulfonyl chloride (*p*-toluenesulfonyl or tosyl chloride) (**11.22**) (Scheme 11.7).[19] Quantitative yields of the products, diaryl sulfones such as **11.23**, were achieved with high reactivity at 30 °C. No reactions occurred with neutral or basic ionic liquids, *i.e.* when the mole fraction of $AlCl_3$ in the binary mixture was equal to or less than 0.5 respectively.

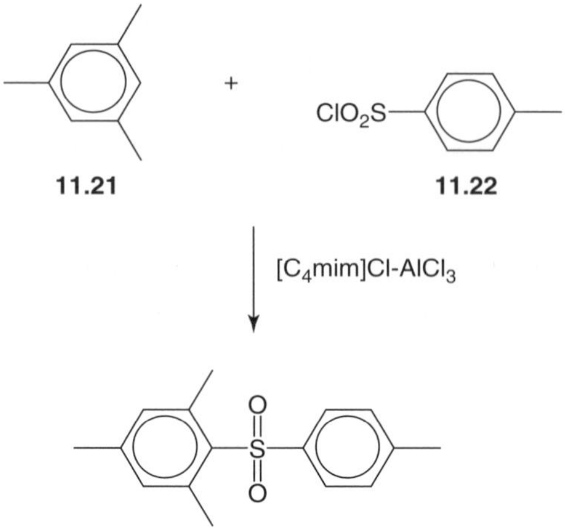

Scheme 11.7   Friedel–Crafts sulfonylation of mesitylene.

Ionic liquids can also be employed for the Friedel–Crafts reactions of aromatic compounds with alkynes. These reactions are known as either alkenylations of arenes or hydroarylations of alkynes.

In 2007, Song and co-workers showed that ionic liquids with $[C_4mim]^+$ cations and very weakly coordinating anions like $[SbF_6]^-$ or $[PF_6]^-$ can be used for these reactions.[20] The team demonstrated that the ionic liquids dramatically enhance the catalytic activities of metal tri-flate-catalysed Friedel–Crafts alkenylations of arenes with alkynes. For example, the Friedel–Crafts reaction of benzene and 1-phenylprop-1-yne (**11.24**) using scandium triflate Sc(OTf)$_3$ as a catalyst was completed far more rapidly and with a much higher yield of the product, 1,1-diphenylprop-1-ene (**11.25**), in $[C_4mim][SbF_6]$ than without the ionic liquid (Scheme 11.8). Furthermore, simple decantation of the product-containing organic layer allowed the catalyst-containing ionic liquid phase to be recovered and reused without significant loss of catalytic activity.

Task-specific ionic liquids have also been prepared for Friedel–Crafts reactions. Qiao and Yokoyame reported, for example, that task-specific Brønsted acidic ionic liquids, such as **11.26**, can used as reaction media and catalysts for the Friedel–Crafts alkylation reaction of benzene, toluene or xylene with styrene (**11.27**) (Scheme 11.9).[21] The ionic liquids were functionalized by covalently tethering an alkane sulfonic acid group to an imidazolium cation. The product-containing upper layer of the biphasic reaction was readily separated by decantation from the lower ionic liquid layer, which was then reused.

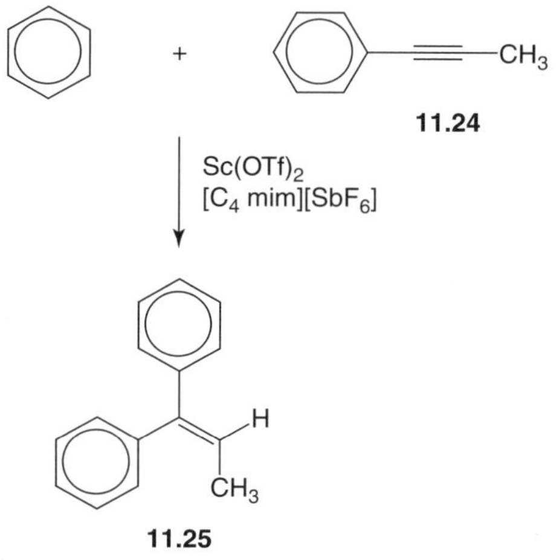

Scheme 11.8   Friedel–Crafts alkenylation of benzene.

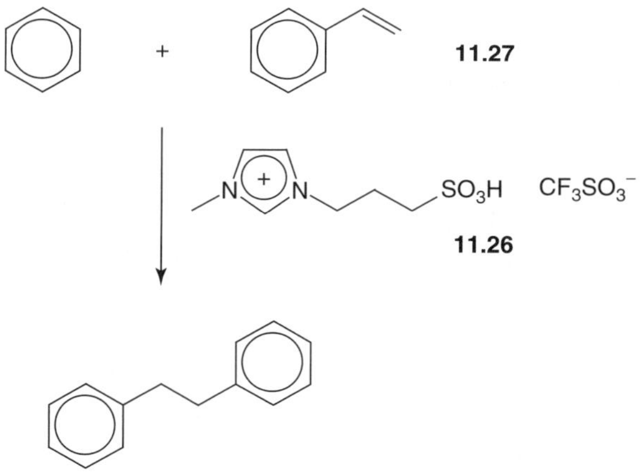

Scheme 11.9   Friedel–Crafts alkylation with a task-specific ionic liquid.

11.6   GRIGNARD REACTION

The Grignard reaction employs alkyl or aryl magnesium halides, RMgX, known as Grignard reagents. These reagents are strong bases and powerful nucleophiles that form carbon–carbon bonds by attacking electrophilic carbon atoms in saturated carbon–heteroatom bonds such as C=O, C=S and C≡N.

Ionic liquids with imidazolium cations cannot be used as solvents for reactions involving strong bases, such as the Grignard reagents, because the cations react with the bases to form imidazolylidenes (*N*-heterocyclic carbenes). In 2005, Clyburne and co-workers showed that phosphonium ionic liquids with $[P_{6\ 6\ 6\ 14}]^+$ cations do not react with strong bases and can therefore be used as solvents for Grignard reactions.[22]

The group carried out, for example, the addition of C_6H_5MgBr to acetone, CH_3COCH_3, to give $C_6H_5C(CH_3)_2OH$ in 82% yield. The reactions were carried out by adding the reactant to a solution of an ionic liquid, such as $[P_{6\ 6\ 6\ 14}]Cl$, and C_6H_5MgBr dissolved in the co-solvent tetrahydrofuran. After the reaction, water and hexane were added to the reaction mixture to form a three-phase system. The products were isolated from the top organic layer. The ionic liquid and aqueous layers were in the middle and on the bottom respectively. After washing with water and hexane and then drying, the ionic liquid was recycled.

In 2006, Chan and co-workers described the preparation of Grignard reagents in ionic liquids.[23] They prepared C_2H_5MgI by adding magnesium metal to ethyl iodide in pure $[C_4py][BF_4]$. However, no reaction occurred when the ethyl iodide was replaced by ethyl bromide.

11.7 HECK REACTION

Of all the numerous types of carbon–carbon coupling reactions that have been carried out in ionic liquids, the Heck reaction is by far the most widely studied. The palladium-catalysed reaction couples alkenes with aryl or vinyl halides to generate substituted alkenes. The palladium(0) catalyst is typically formed using a base like triethylamine to reduce a mixture of a palladium salt, such as palladium(II) acetate or palladium(II) chloride, and triphenylphosphine ligands. The reaction works well with alkenes possessing an electron-withdrawing substituent, *e.g.* nitrile.

One of the problems of the Heck reaction is recovery of the palladium catalyst at the end of the reaction. In 1999, Earle and co-workers showed that ionic liquids can be used to recycle the palladium catalyst used for the Heck reactions of an aromatic halide or anhydride with an alkene to give an aryl alkene.[24] The team coupled various aromatic compounds with ethyl and butyl acrylates in $[C_4mim][PF_6]$, $[C_6py]Cl$ and related ionic liquids, using triethylamine or $NaHCO_3$ as the base and $Pd(OAc)_2$ with phosphine ligands as the catalyst precursor.

In multiphasic systems, the palladium catalysts dissolve in the ionic liquids in preference to water or alkane solvents. Earle's team demonstrated, for example, that the reaction of iodobenzene (**11.28**) and ethyl acrylate (**11.29**) to yield ethyl cinnamate (**11.30**), can be run in a triphasic system consisting of a lower ionic liquid layer that contains the catalyst,

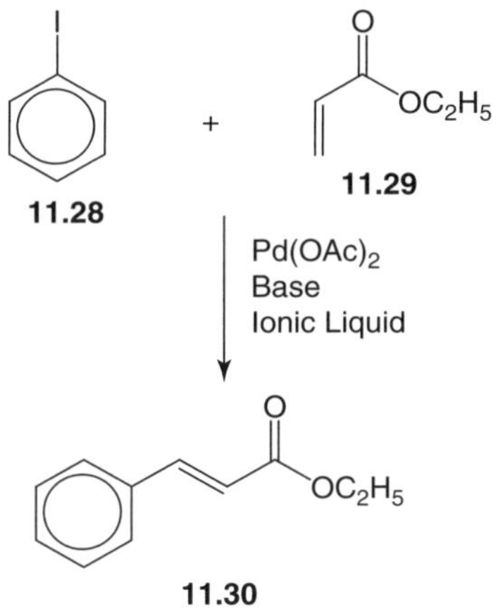

Scheme 11.10 Heck reaction of iodobenzene and ethyl acrylate.

an aqueous middle layer and a cyclohexane top layer (Scheme 11.10). The aqueous layer removes salt by-products from the reaction and the cyclohexane layer extracts the cinnamate. The catalyst-containing ionic liquid layer can then be reused repeatedly without loss of catalytic activity.

In 2000, Xiao and co-workers reported that the activity and stability of the palladium catalyst depends on the choice of ionic liquids.[25] The group investigated the Heck olefination reactions of various aryl halides with acrylates and styrene and showed that the reactions were much more efficient in [C$_4$mim]Br than in [C$_4$mim][BF$_4$]. They attributed the improved efficiency to the formation of a *N*-heterocyclic carbene complexes of palladium such as **11.31** in [C$_4$mim]Br. This complex proved to be catalytically active for carbon–carbon single bond formation in [C$_4$mim]Br but reacted with [C$_4$mim][BF$_4$], thereby losing some of its activity.

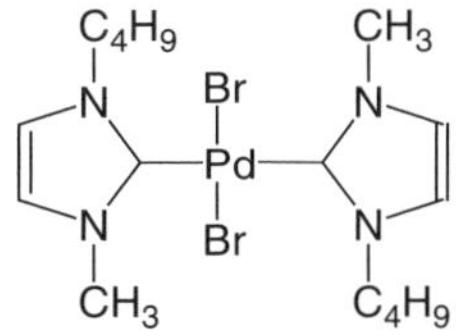

11.31

The following year, Xiao's group showed that the Heck reaction of aromatic halides with electron-rich olefins proceeds with "high efficiency and remarkable regioselectivity" in [C$_4$mim][BF$_4$].[26] The team carried out the arylation of the electron-rich olefin butyl vinyl ether using aryl iodides and bromides as arylating agents and a palladium catalyst with phosphine ligands.

The arylations led almost exclusively to substitution at the olefinic carbon next to the oxygen atom of C_4H_9-O-CH=CH$_2$, *i.e.* α-substitution of the butyl vinyl ether. In organic molecular solvents such as dimethylformamide, the Heck arylations resulted in both α- and β-substitution, typically in the ratio 60 : 40.

The chemists attributed the impact of the ionic liquid on the regioselectivity to the "ionic liquid effect", *i.e.* the impact of the unique ionic environment of the reactants on the course of chemical reaction.

Phosphine ligands can, in some instances, oxidize in the presence of air and moisture to form phosphine oxide species that poison the palladium catalyst. In 2003, Park and Alper described Heck reactions that can be carried out in ionic liquids without the use of phosphine ligands for the palladium catalyst.[27] The two chemists showed that a palladium(II) complex with the bisimidazole ligand **11.32** is stable to air and moisture and is an effective catalyst for coupling iodobenzene with *n*-butyl acrylate, CH_2=CH-CO$_2$C$_4$H$_9$, in the ionic liquid [C$_4$mim][PF$_6$]. They achieved quantitative yields of the cinnamate product and showed that the catalyst could be recycled five times without loss of catalytic activity.

11.32

11.8   KNOEVENAGEL CONDENSATION AND ROBINSON ANNULATION

The Knoevenagel condensation is a nucleophilic addition reaction in which an aldehyde or ketone combines with an active methylene compound with two electron-withdrawing groups. The reaction results in the loss of a molecule of water and the formation of an α,β-unsaturated dicarbonyl or related compound. The reaction is an aldol-type condensation (Section 10.2) catalysed by a weak base, normally a basic amine. The base removes an acidic hydrogen atom from the active methylene compound to generate a nucleophilic carbanion.

The Robinson annulation is a ring-forming reaction of a ketone, usually a cyclic ketone, with an α,β-unsaturated ketone that produces a substituted cyclohexenone compound. The reaction involves a Michael addition (Section 11.10) followed by an aldol condensation.

Davis and co-workers employed [C₆mim][PF₆] as a solvent for both reactions.[28] For the Knoevenagel condensation, they used glycine (**11.33**) as the base for the reaction of malonitrile (propane-1,3-dinitrile, **11.34**) and benzaldehyde (**11.35**), both of which dissolve readily in the ionic liquid (Scheme 11.11). The product, 1,1-dicyano-2-phenylethene (**11.36**), was extracted in "excellent" yield using toluene.

For the Robinson annulation, the group used sodium hydroxide as the base for the reaction of ethyl acetoacetate (**11.37**) and chalcone (1,3-diphenyl-2-propen-1-one, **11.38**) (Scheme 11.12). The product, 6-ethoxycarbonyl-3,5-diphenylcyclohex-2-enone (**11.39**), was obtained in "modest" yield following neutralization of the reaction mixture, extraction with toluene, and silica gel chromatography.

After both reactions, the ionic liquid was recycled for several runs without loss of yield. Several other investigations of the Knoevenagel condensation in ionic liquids have been carried out following the early

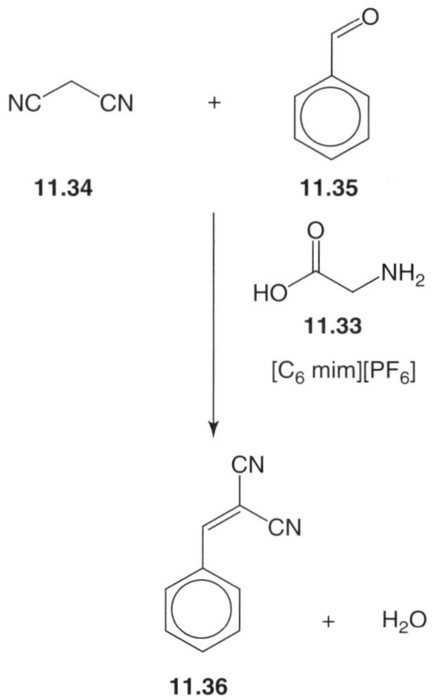

Scheme 11.11 Knoevenagel condensation.

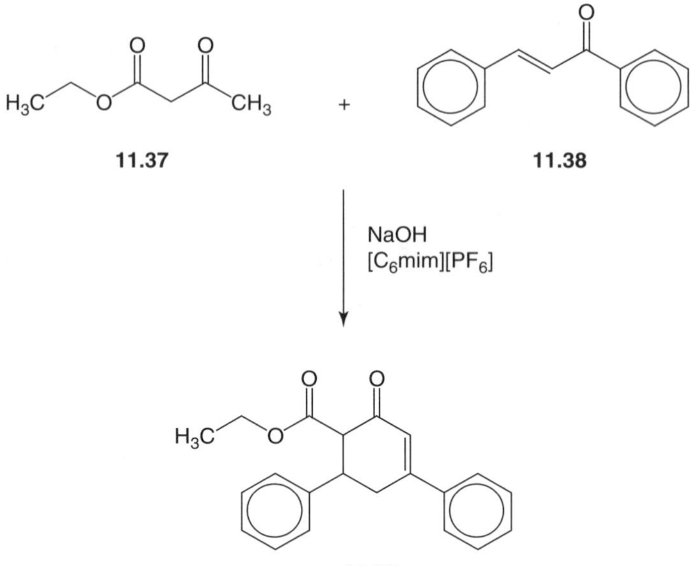

11.37   11.38

NaOH
[C$_6$mim][PF$_6$]

11.39

Scheme 11.12   Robinson annulation.

work by Davis's group. One example is a study by Ranu and Jana of the condensation of a wide range of aliphatic and aromatic aldehydes and ketones with active methylene compounds such as malonitrile using the basic task-specific ionic liquid [C$_4$mim]OH as a catalyst and reaction medium.[29]

The reactions proceeded rapidly at room temperature, leading to high isolated yields of products. The two chemists showed, for example, that the condensation of a hydroxybenzaldehyde, such as salicylaldehyde (**11.40**), with diethyl malonate (**11.41**), which is difficult to achieve using conventional reagents, proceeded without difficulty in the ionic liquid, resulting in high yields of bicyclic compounds known as coumarins, *e.g.* **11.42** (Scheme 11.13).

11.9   MANNICH REACTION

Kabalka and co-workers have shown that the highly polarizable nature of ionic liquids can be used to catalyse a variation of the Mannich reaction known as the Petasis reaction.[30] The Mannich reaction is a three-component condensation reaction of an aldehyde, a primary or secondary amine or ammonia, and a nucleophilic compound, usually an

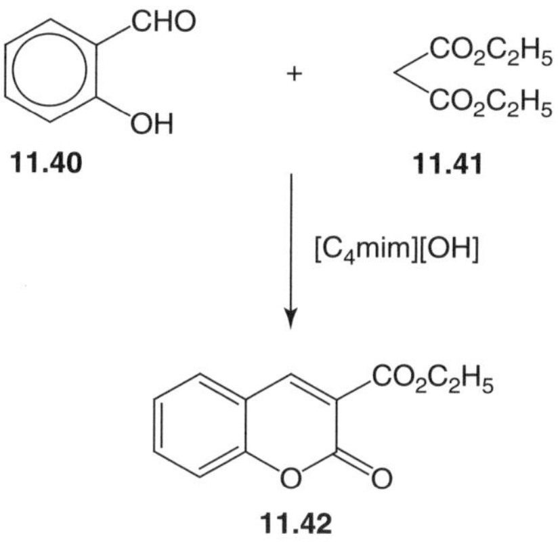

Scheme 11.13 Knoevenagel condensation with a basic task-specific ionic liquid.

aldehyde or ketone, containing an acidic hydrogen atom. The reaction, which is typically carried out under acidic conditions, generates substituted β-amino carbonyl compounds. These aminoalkylated products are known as Mannich bases. The Petasis variation employs a Lewis acidic organoborane as the nucleophilic component.

Kabalka's team showed that ionic liquids like [C₄mim][BF₄] are suitable solvents for the reaction of potassium alkynyltrifluoroborates with amines and salicylaldehydes, such as $C_4H_9C{\equiv}CBF_3K$ with $(C_6H_5CH_2)_2NH$ and **11.40**, respectively, in the presence of benzoic acid. The reaction produced highly functionalized propargylamines, *e.g.* **11.43**, in good yields.

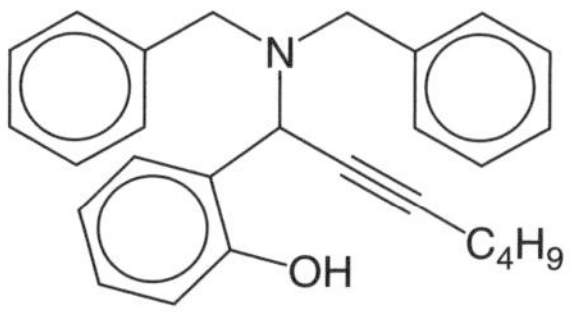

11.43

11.10 MICHAEL ADDITION

The Michael addition is a carbon–carbon single bond forming reaction in which a nucleophilic carbanion undergoes conjugate addition to an α,β-unsaturated carbonyl compound. The carbon nucleophile is known as the Michael donor, the carbonyl compound the Michael acceptor, and the product the Michael adduct.

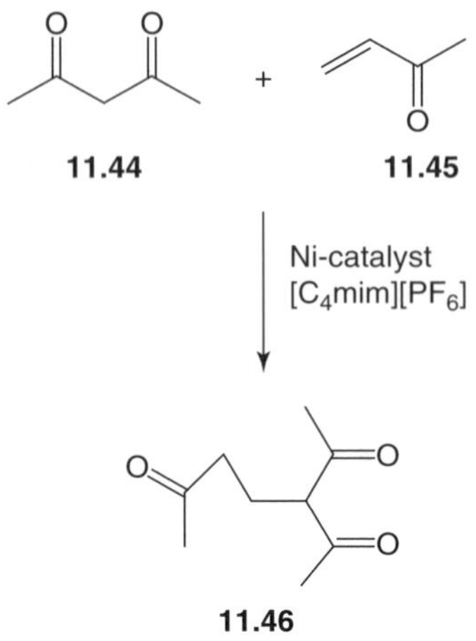

Scheme 11.14 Michael addition of pentane-2,4-dione to methyl vinyl ketone.

Various Michael additions have been carried out in ionic liquids. In 2002, for example, Nobile and co-workers described the use of [C$_4$mim][PF$_6$] as a solvent for the metal-catalysed Michael addition of pentane-2,4-dione (**11.44**) to the Michael acceptor methyl vinyl ketone (**11.45**).[31] The team tested three catalysts and found that nickel(II) acetylacetonate, Ni(CH$_3$COCHCOCH$_3$)$_2$, was "outstanding" in terms of activity, yielding 94% of the Michael adduct, 3-acetyl-2,6-heptane-dione (**11.46**), with more than 98% selectivity (Scheme 11.14). The product was removed by distillation and the catalyst and ionic liquid solution recycled several times.

Ionic liquids can act not only as solvents for Michael additions but also as catalysts. For example, a range of task-specific chiral ionic liquids have been developed as asymmetric catalysts for the Michael addition of various Michael ketone and aldehyde donors to nitroolefin Michael acceptors.[32] The chiral ionic liquids were pyrrolidine–ionic liquid conjugates, such as **11.47**, synthesized from the starting materials imidazole and L-proline.

The asymmetric Michael additions were carried out at room temperature in neat mixtures using trifluoroacetic acid as a co-catalyst (Scheme 11.15). The reaction of cyclohexanone (**11.48**), the Michael donor, with *trans-β*-nitrostyrene (**11.49**) as the Michael acceptor, for example, yielded a product with 99% enantioselectivity and 99 : 1 *syn/anti* diastereoselectivity. The prefixes *syn-* and *anti-* refer to diastereoisomers formed by the addition of substituents to the same side or to the opposite sides, respectively, of a double bond.

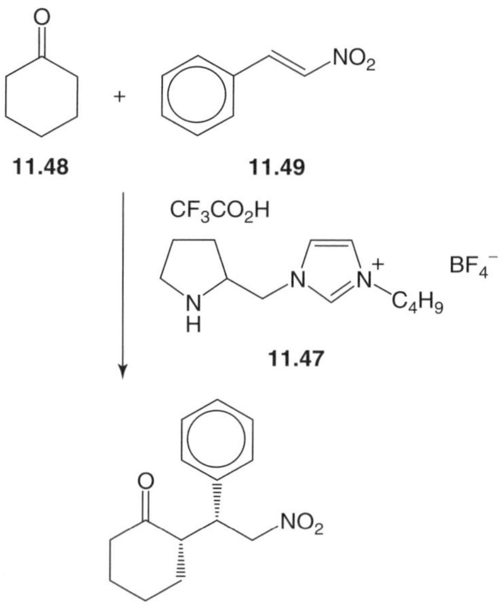

Scheme 11.15 Asymmetric Michael addition with a task-specific ionic liquid.

The addition of amines to α,β-unsaturated carbonyl compounds, known as the aza-Michael addition, has also been carried out in ionic liquids. Kantam and colleagues showed that copper(II) acetylacetonate, Cu(acac)$_2$, immobilized in [C$_4$mim][PF$_6$] or [C$_4$mim][BF$_4$] catalyses the reactions "with great alacrity and excellent yields" of β-amino ketones.[33] The Cu(acac)$_2$-containing ionic liquid phase was recovered and reused several times without loss of catalytic activity.

11.11 SONOGASHIRA COUPLING

Sonogashira reactions typically use a palladium-phosphine complex, a co-catalyst, which is usually a copper(I) halide, and a base to couple aryl or vinyl halides to a compound with a terminal alkyne.

In 2004, Park and Alper showed that the reaction can be carried out in [C$_4$mim][PF$_6$] in the absence of a copper salt or phosphine.[34] The chemists used a palladium-imidazole catalyst and an amine base, *e.g.* piperidine (**11.50**), to couple iodobenzene (**11.51**) and related compounds with phenylacetylene (**11.52**) and alkyl acetylenes (Scheme 11.16). The reactions proved to be "efficient." The products were separated from the catalyst by extraction with ethyl ether. The ionic liquid layers were then washed with water to remove the amine salts and reused.

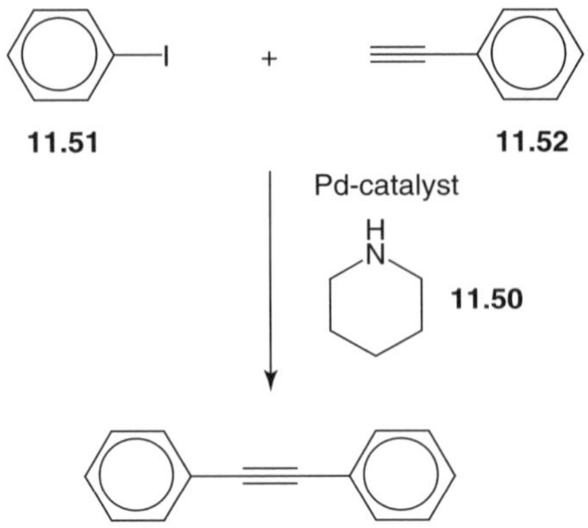

Scheme 11.16   Sonogashira coupling.

11.12   STILLE COUPLING

The Stille reaction is a palladium(0)-catalysed carbon–carbon single bond forming reaction that couples an organotin compound with an organic halide. The first example of the reaction in an ionic liquid was described by Handy and Zhang in 2001.[35] The two chemists used various palladium catalyst systems to couple organic halides with a stannane such as phenyltributyltin, $C_6H_5Sn(C_4H_9)_3$. For instance, they prepared biphenyl **(11.53)** in 90% yield by coupling bromobenzene with phenyltributyltin in [C_4mim][PF_6], using the palladium-triphenylphosphine precursor, $Pd[(C_6H_5)_3P]_4$, as a source of the palladium(0) catalyst. The ionic liquid and catalyst systems were used several times with little loss of activity.

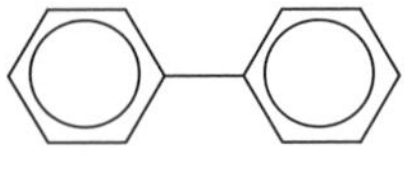

11.53

Chiappe, Dyson and co-workers showed that task-specific ionic liquids with cations bearing nitrile functional groups and nitrogen-containing anions have a pronounced impact on the efficiency of Stille reactions.[36] For example, the yield of styrene, $C_6H_5CH=CH_2$, obtained by the transfer of a vinyl group from tributylvinylstannane, $(C_4H_9)_3SnCH=CH_2$, to bromobenzene in [C_3CNmim][NTf_2] **(11.54)** using palladium(II) acetate as a catalyst precursor was almost double that obtained for the same reaction in [C_4mim][NTf_2]. The authors attributed the increased efficiency to the interaction of the cation's nitrile group with the palladium catalyst. The ionic liquid, which has a nitrogen-containing anion, also proved "far superior" as a solvent to [C$_3$CNmim][BF$_4$] for this reaction.

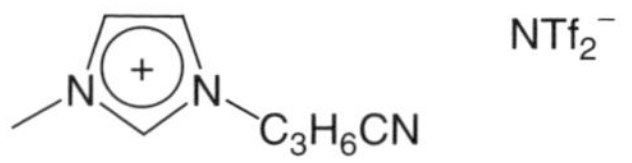

$$NTf_2^-$$

11.54

11.13 SUZUKI CROSS-COUPLING

The Suzuki cross-coupling reaction is another palladium-catalysed carbon–carbon single bond forming reaction that has been widely studied in ionic liquids. The reaction couples an aryl or vinyl boronic acid with an aryl or vinyl halide. It is commonly used to synthesize biaryl compounds.

In 2000, Welton and co-workers reported the first examples of Suzuki cross-coupling reactions in an ionic liquid.[37] By using suspensions of the palladium-phosphine catalyst, Pd[(C$_6$H$_5$)$_3$P]$_4$, in [C$_4$mim][BF$_4$], the team achieved "unprecedented reactivities in addition to easy product isolation and catalyst recycling." In a typical reaction, 4-bromotoluene (**11.55**) was coupled with phenylboronic acid (**11.56**) to produce 4-methylbiphenyl (**11.57**) (Scheme 11.17).

Welton's team subsequently showed that palladium imidazolylidene (*N*-heterocyclic carbene) complexes such as **11.58** are produced as stable catalytic forms during these reactions.[38]

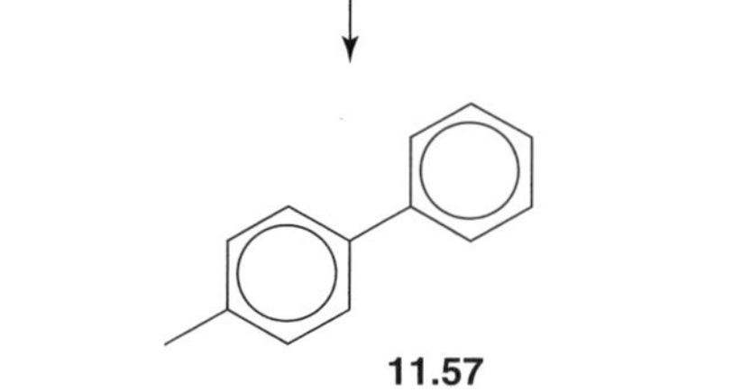

Scheme 11.17 Suzuki cross-coupling.

11.58   Ph =

The team also showed that Suzuki cross-coupling can be carried out in dialkylimidazolium room-temperature ionic liquids in the absence of phosphine ligands by using palladium-imidazole complexes as catalyst precursors.[39] The complexes were generated by adding $(CH_3CN)_2PdCl_2$ and an imidazole such as 1-methylimidazole to the ionic liquid. The chemists demonstrated that the activity and stability of the catalyst in the ionic liquid depended on the nature of the imidazole ligands and the ionic liquid's anions and cations.

11.14   SWERN OXIDATION

Swern oxidation is the oxidation of primary or secondary alcohols to aldehydes or ketones using dimethyl sulfoxide (DMSO) as the oxidant. The DMSO is activated by either trifluoroacetic anhydride or oxalyl chloride, $(COCl)_2$. A tertiary amine, typically triethylamine, $(C_2H_5)_3N$, is added to decompose the alkoxysulfonium salt that is formed during the reaction. The salt decomposes to dimethyl sulfide (DMS) and the ketone or aldehyde.

DMS, like many other organosulfur compounds, is volatile, toxic and has an unpleasant odour. Its generation is a drawback of the reaction. In 2006, He and Chan, described how the problem can be avoided by grafting sulfoxides onto a dimethylimidazolium ionic liquid (**11.59**), thereby rendering the DMSO and DMF non-volatile and odorless.[40] They employed the anchored sulfoxides for the Swern oxidation of 12 alcohols. $(COCl)_2$ was used to activate the sulfoxides and $(C_2H_5)_3N$ used as the base. Good yields of the products were obtained. For example, the oxidation of benzhydrol (**11.60**) resulted in a 90% yield of benzophenone (**11.61**) (Scheme 11.18). The sulfide by-product **11.62**, formed during the process, remained anchored to the ionic liquid. Addition of periodic acid, H_5IO_6, converted the anchored sulfide back into the anchored sulfoxide, which could be recovered and reused.

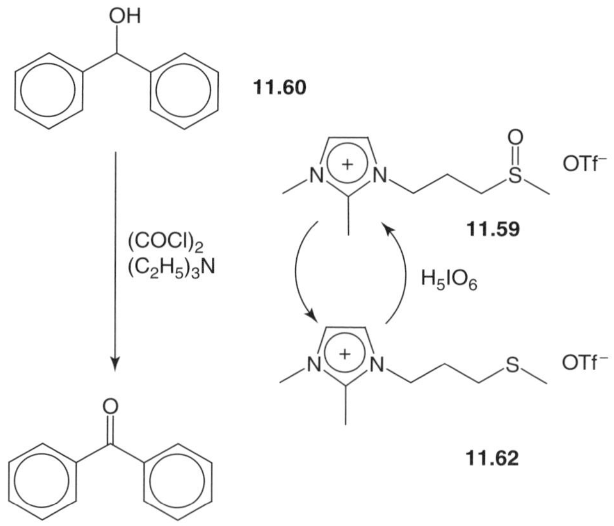

Scheme 11.18   Swern oxidation.

11.15   ULLMANN COUPLING

The Ullmann reaction conventionally couples two aryl halides in the presence of a copper catalyst to form biaryl compounds.

Bao and co-workers used an Ullmann-type reaction to couple vinyl bromides with imidazoles in [C$_4$mim][BF$_4$] and other ionic liquids using copper(II) iodide, CuI$_2$, as the catalyst and L-proline as the ligand.[41] The reaction of β-bromostyrene (**11.63**) and imidazole (**11.64**) in [C$_4$mim][BF$_4$] in the presence of K$_2$CO$_3$ produced the coupled product *N*-styrylimidazole (**11.65**) in 87% yield (Scheme 11.19). The method proved to be "mild and efficient" and the ionic liquids were "excellent media" for the coupling reactions, according to Bao's team.

The Ullmann reaction generates significant amounts of CuI$_2$ waste. In 2006, Rothenberg and colleagues showed that Ullmann reactions can be carried out electrochemically in ionic liquids at room temperature without the use of copper salts.[42] The process involves the reduction and coupling of aryl bromides and iodides in [C$_8$mim][BF$_4$] in a cell containing a palladium anode and platinum cathode. The palladium is oxidized at the anode, forming Pd^{2+} ions that are then reduced at the cathode. The reduction leads to the formation of Pd(0) nanoparticles that catalyse the electroreductive homocoupling of haloarenes (Scheme 11.20).

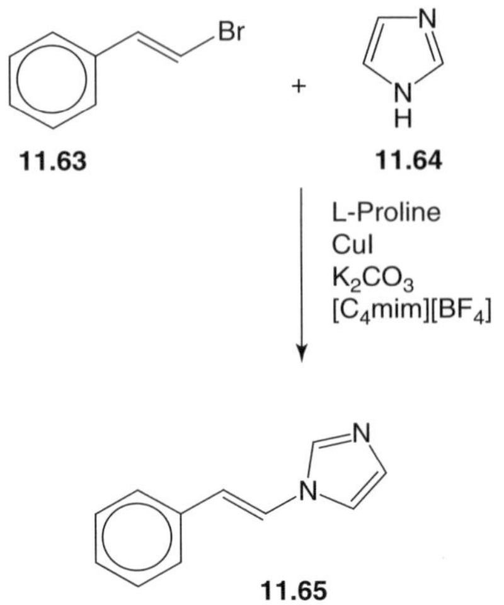

Scheme 11.19 Ullmann-type coupling.

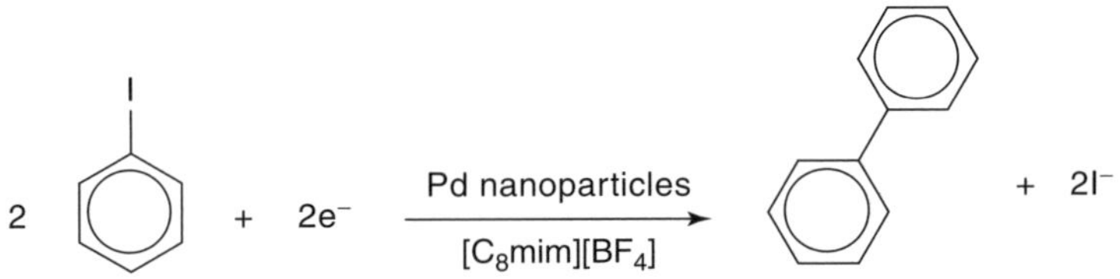

Scheme 11.20 Electroreductive Pd-catalysed Ullmann reaction.

11.16 WACKER OXIDATION

The Wacker oxidation is the palladium(II)-catalysed oxidation of olefins to the corresponding ketones. It is usually carried out homogeneously in water using oxygen or hydrogen peroxide as the oxidant.

In 2002, Varma and co-workers demonstrated that the Wacker oxidation of styrene to acetophenone can be carried out in the presence of [C$_4$mim][PF$_6$] or [C$_4$mim][BF$_4$] using PdCl$_2$ as a catalyst and hydrogen peroxide as the oxidant.[43] They showed that the ionic liquids enhance the conversion of styrene, the selectivity for acetophenone and the activity of the catalyst compared with Wacker oxidations of styrene using other methods.

11.17 WITTIG REACTION

The Wittig reaction is a carbon–carbon double bond forming reaction that couples a carbonyl compound, an aldehyde or a ketone, with a

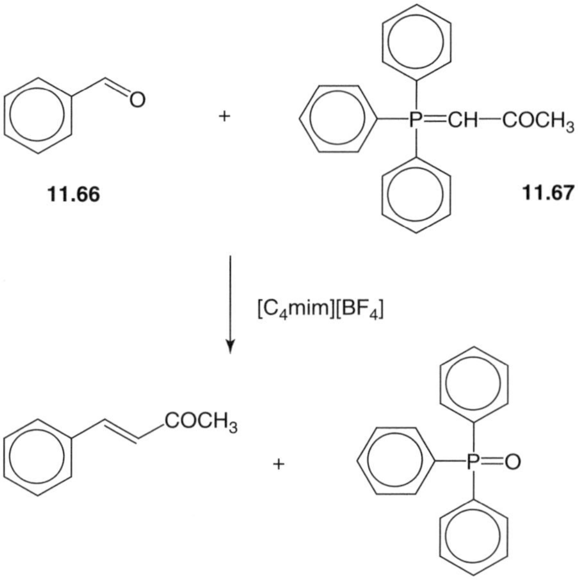

Scheme 11.21 Wittig reaction.

phosphorane (also known as a phosphorus ylide). The products of the Wittig reaction are an alkene and triphenylphosphine oxide, $(C_6H_5)_3PO$. The alkene is conventionally separated from the oxide by crystallization or chromatography.

An ylide is an internal salt in which a moiety containing a heteroatom, such as phosphorus, is cationic. A classic example is $(C_6H_5)_3P^+-CH_2:^-$. An ylide with at least one strong electron-withdrawing group, such as $-COCH_3$, that stabilizes the negative charge on the carbon atom attached to the heteroatom is known as a stabilized ylide.

In 2000, Le Boulaire and Grée showed that the alkenes can be easily separated by using an ionic liquid as a solvent for the reaction.[44] They employed $[C_4mim][BF_4]$ as a solvent for the reactions of benzaldehyde (**11.66**) with stabilized ylides such as **11.67** (Scheme 11.21). The alkenes were extracted from the reaction mixtures with diethyl ether. Toluene was then used to extract $(C_6H_5)_3PO$ and unreacted ylide, leaving the ionic liquid for recycling.

REFERENCES

1. R. Bernini, A. Coratti, G. Fabrizi and A. Goggiamani, *Tetrahedron Lett.*, 2003, **44**, 8991.

2. J. N. Rosa, C. A. M. Afonso and A. G. Santos, *Tetrahedron*, 2001, **57**, 4189.
3. B. Pégot, G. Vo-Thanh, D. Gori and A. Loupy, *Tetrahedron Lett.*, 2004, **45**, 6425.
4. J. Peng and Y. Deng, *Tetrahedron Lett.*, 2001, **42**, 403.
5. D. A. Jaeger and C. E. Trucker, *Tetrahedron Lett.*, 1989, **30**, 1785.
6. J. Howarth, K. Hanlon, D. Fayne and P. McCormac, *Tetrahedron Lett.*, 1997, **38**, 3097.
7. C. W. Lee, *Tetrahedron Lett.*, 1999, **40**, 2461.
8. M. J. Earle, P. B. McCormac and K. R. Seddon, *Green Chem.*, 1999, **1**, 23.
9. S. Tiwari and A. Kumar, *Angew Chem. Int. Ed.*, 2006, **45**, 4824.
10. C. E. Song, W. H. Shim, E. J. Roh, S. Lee and J. H. Choi, *Chem. Commun.*, 2001, 1122.
11. J. A. Boon, J. A. Levisky, J. L. Pflug and J. S. Wilkes, *J. Org. Chem.*, 1986, **51**, 480.
12. S. Csihony, H. Mehdi and I. T. Horváth, *Green Chem.*, 2001, **3**, 307.
13. A. K. Abdul-Sada, M. P. Atkins, B. Ellis, P. K. G. Hodgson, M. L. M Morgan and K. R. Seddon, *World Pat.*, WO 95/21806 (17 August 1995).
14. M. P. Atkins, C. Bowlas, B. Ellis, F. Hubert, A. Rubatto and P. Wasserscheid, in *Green Industrial Applications of Ionic Liquids*, ed. R. D. Rogers, K. R. Seddon and S. Volkov, NATO Science Series, Kluwer Academic Publishers, Dordrecht, 2002, p. 49.
15. C. J. Adams, M. J. Earle, G. Roberts and K. R. Seddon, *Chem. Commun.*, 1998, 2097.
16. J. Ross and J. Xiao, *Green Chem.*, 2002, **4**, 129.
17. M. J. Earle, U. Hakala, B. J. McAuley, M. Nieuwenhuyzen, A. Ramani and K. R. Seddon, *Chem. Commun.*, 2004, 1368.
18. Z.-W. Wang and L.-S. Wang, *Appl. Catal., A*, 2004, **262**, 101.
19. S. J. Nara, J. R. Harjani and M. M. Salunkhe, *J. Org. Chem.*, 2001, **66**, 8616.
20. M. Y. Yoon, J. J. Kim, D. S. Choi, U. S. Shin, J. Y. Lee and C. E. Song, *Adv. Synth. Catal.*, 2007, **349**, 1725.
21. K. Qiao and C. Yokoyama, *Chem. Lett.*, 2004, **33**, 472.
22. T. Ramnial, D. D. Ino and J. A. C. Clyburne, *Chem. Commun.*, 2005, 325.
23. M. C. Law, K.-Y. Wong and T. H. Chan, *Chem. Commun.*, 2006, 2457.
24. A. J. Carmichael, M. J. Earle, J. D. Holbrey, P. B. McCormac and K. R. Seddon, *Org. Lett.*, 1999, **1**, 997.
25. L. Xu, W. Chen and J. Xiao, *Organometallics*, 2000, **19**, 1123.

26. L. Xu, W. Chen, J. Ross and J. Xiao, *Org. Lett.*, 2001, **3**, 295.
27. S. B. Park and H. Alper, *Org. Lett.*, 2003, **5**, 3209.
28. D. W. Morrison, D. C. Forbes and J. H. Davis Jr, *Tetrahedron Lett.*, 2001, **42**, 6053.
29. B. C. Ranu and R. Jana, *Eur. J. Org. Chem.*, 2006, 3767.
30. G. W. Kabalka, B. Venkataiah and G. Dong, *Tetrahedron Lett.*, 2004, **45**, 729.
31. M. M. Dell'Anna, V. Gallo, P. Mastrorilli, C. F. Nobile, G. Romanazzi and G. P. Suranna, *Chem. Commun.*, 2002, 434.
32. S. Luo, X. Mi, L. Zhang, S. Liu, H. Xu and J.-P. Cheng, *Angew. Chem. Int. Ed.*, 2006, **45**, 3093.
33. M. L. Kantam, V. Neeraja, B. Kavita, B. Neelima, M. K. Chaudhuri and S. Hussain, *Adv. Synth. Catal.*, 2005, **347**, 763.
34. S. B. Park and H. Alper, *Chem. Commun.*, 2004, 1306.
35. S. T. Handy and X. Zhang, *Org. Lett.*, 2001, **3**, 233.
36. C. Chiappe, D. Pieraccini, D. Zhao, Z. Fei and P. J. Dyson, *Adv. Synth. Catal.*, 2006, **348**, 68.
37. C. J. Matthews, P. J. Smith and T. Welton, *Chem. Commun.*, 2000, 1249.
38. P. J. Smith and T. Welton, in *Ionic Liquids: Industrial Applications to Green Chemistry*, ed. R. D. Rogers and K. R. Seddon, ACS Symposium Series 818, American Chemical Society, Washington DC, 2002, p. 310.
39. C. J. Mathews, P. J. Smith and T. Welton, *J. Mol. Catal. A: Chem.*, 2004, **214**, 27.
40. X. He and T. H. Chan, *Tetrahedron*, 2006, **62**, 3389.
41. Z. Wang, W. Bao and Y. Jiang, *Chem. Commun.*, 2005, 2849.
42. L. D. Pachón, C. J. Elsevier and G. Rothenberg, *Adv. Synth. Catal.*, 2006, **348**, 1705.
43. V. V. Namboodiri, R. S. Varma, E. Sahle-Demessie and U. R. Pillai, *Green Chem.*, 2002, **4**, 170.
44. V. Le Boulaire and R. Grée, *Chem. Commun.*, 2000, 2195.

CHAPTER 12
Biotechnology

12.1 INTRODUCTION

Biotechnology, in its broadest sense, is the technological use of biological organisms and their components for the industrial and commercial production of agricultural crops, chemicals, energy, pharmaceuticals and other products. The use of ionic liquids for such biotechnological applications is in its infancy. Much of the research to date has focused on enzyme catalysis. There has also been interest in the use of ionic liquids to convert biomass into useful chemical products and biofuels.

12.2 CARBOHYDRATE SOLUBILITY

The use of biomass to replace petroleum feedstocks requires efficient methods for converting naturally-occurring carbohydrates into industrially useful chemical products. There is now a growing amount of evidence that ionic liquids may possibly prove useful for such conversions. In particular, it is well established that ionic liquids can be selected or designed to dissolve a wide range of monosaccharides, disaccharides, trisaccharides and polysaccharides.

One of the earliest papers alluding to the solubility of carbohydrates in ionic liquids appeared in 2001. The paper reported a range of ionic liquids that dissolve glucose.[1] One of them, 1-(2-methoxyethyl)-3-methylimidazolium tetrafluoroborate (**12.1**), dissolves 100 times more glucose than does acetone or tetrahydrofuran.

An Introduction to Ionic Liquids
By Michael Freemantle
© Michael Freemantle 2010
Published by the Royal Society of Chemistry, www.rsc.org

$$BF_4^-$$

H_3C—N(+)N—CH_2CH_2OCH_3

12.1

In the same year, Kimizuka and Nakashima discovered that similar ionic liquids with ether-containing cations can dissolve not only glucose but also polysaccharides such as α-cyclodextrin, amylase and agarose.[2] The macrocyclic carbohydrate α-cyclodextrin, for example, is soluble in 1-methoxyethyl-3-methylimidazolium bromide. The two researchers attributed the solubility of carbohydrates in these "sugar-philic" ionic liquids to the cation's ether linkage, which possibly acts as a hydrogen-bond acceptor for the carbohydrate's hydroxyl groups. Kimizuka and Nakashima also showed that glucose oxidase, a glycosylated protein, is soluble in ionic liquids with ether-containing cations.

Anions can also play a key role in determining the solubility of carbohydrates in ionic liquids. For example, monosaccharides such as glucose, disaccharides such as sucrose and lactose, trisaccharides such as raffinose, and polysaccharides such as cyclodextrin are all soluble in the imidazolium dicyanamide ionic liquids [C_2mim][dca] and [C_4mim][dca] whereas their solubility in ionic liquids with the same cations but other anions is low.[3]

In 2002, Rogers and co-workers reported that imidazolium ionic liquids with anions that are strong hydrogen bond acceptors dissolve cellulose pulp.[4] Chloride-containing ionic liquids such as [C_4mim]Cl are most effective, especially when heated in a microwave oven. Ionic liquids with non-coordinating anions such as [BF_4]$^-$ and [PF_6]$^-$ do not dissolve cellulose pulp, even under microwave heating.

The authors suggested that the solubility of the cellulose in the chloride-containing ionic liquids results from the breaking of the intramolecular hydrogen-bonding network in the cellulose and the formation of hydrogen bonds between the chloride anions and the hydroxyl groups of the cellulose.

Cellulose not only rapidly dissolves in the imidazolium chloride ionic liquids it can also be readily regenerated by adding water to the ionic liquid–cellulose solution. The cellulose precipitates out and can be coagulated into fibres, membranes, beads and other forms using conventional extrusion spinning or forming techniques.

Additives can also be dissolved or suspended in ionic liquid–cellulose solutions to make cellulose composites with a range of properties. For example, magnetically active cellulose fibres have been spun from a dispersion of magnetite (Fe_3O_4) particles in a [C_2mim]Cl-cellulose solution.[5] Magnetic papers and fibres consisting of cellulose-magnetite composites are used in a range of applications such as magnetic filters and security paper.

In a publication with the phrase "'green' solvents meet green bananas" in its title, Rogers and colleagues demonstrated that [C$_4$mim]Cl is capable of completely dissolving the pulps of peeled green, green/yellow, yellow and yellow/brown bananas. The solutions enabled the changes in the carbohydrate composition of the fruit at various stages of ripening to be analysed directly by ^{13}C nuclear magnetic resonance (NMR) spectroscopy.[6]

The same ionic liquid has also been shown to partially dissolve woods of varying hardness, *e.g.* pine, poplar, eucalyptus and oak.[7] Cellulose that is virtually free from lignin and hemicellulose, which are the two other major components of wood, can be recovered from the ionic liquid wood extracts by adding a precipitating solvent such as water, ethanol or acetone.

12.3 BIOMASS CONVERSION

Lignocellulosic biomass, also known as plant biomass, is the most important bio-renewable resource on Earth. The conversion of this sustainable feedstock into biofuels and chemicals such as ethanol, by fermentation for example, can help overcome our reliance on the non-renewable fossil fuels such as oil and coal.

The three main components of lignocellulosic biomass are cellulose, hemicelluloses and lignin. Cellulose consists of long chains of glucose units and is the most abundant natural organic polymer on earth. Hemicelluloses are branched polysaccharides consisting of a range of monosaccharide monomers of which xylose is the most predominant. Lignin is a complex crosslinked polymer consisting of phenolic components.

One of the most attractive bioprocesses for converting plant biomass into fuels and chemicals is the bacterial fermentation of biomass carbohydrates to acetone (A), butanol (B) and ethanol (E). The process is known as ABE fermentation. Ethanol is readily removed from the fermentation broth by distillation. However, distillation of the broth to recover butanol, an important industrial chemical and potential biofuel, is not economical. The most promising technology for recovering the alcohol is pervaporation. In this process the alcohol permeates through a membrane and then evaporates. A liquid feed is applied to one side of the membrane and a vacuum to the other side.

In 2001, Fadeev and Meagher examined the potential of ionic liquids for the extractive pervaporation of butanol from fermentation broth.[8]

They demonstrated that [C₄mim][PF₆] and [C₈mim][PF₆] extract butanol with some water from aqueous solutions. However, the pervaporation step did not enhance the separation of butanol from the water. The two researchers suggested that the recovery of butanol from the ionic liquid by distillation might be more economical. Nevertheless, the work provided evidence that ionic liquids are potentially useful for extractive separations in biotechnology.

A related paper, published in 2004 by Matsumoto and colleagues, focused on the recovery of lactic acid from fermentation broth.[9] Lactic acid is used to make polylactic acid, a biodegradable polymer. The work showed that the ability of imidazolium ionic liquids to extract the acid is very low. However, the lactic acid extraction behaviour of imidazolium ionic liquids containing the extractant tri-*n*-butylphosphate is similar to that of solutions of conventional organic solvents with the same extractant.

Recent work by Wang and Zhao indicates that acid-catalysed hydrolysis of lignocellulosic materials dissolved in an imidazolium ionic liquid may be a cost-effective method for breaking down lignocellulose into glucose, xylose and other monosaccharides.[10] In a series of experiments, corn stalk, rice straw, pine wood and bagasse were added to mixtures of [C₄mim]Cl and hydrochloric acid. The process produced good yields of the monosaccharides. Similar results were obtained for other imidazolium ionic liquids such as [C₄mim]Br and 1-allyl-3-methylimidazolium chloride. The researchers suggested in their paper that dissolving the lignocellulosic materials in the ionic liquids promotes the dispersion of most of the cellulose and hemicellulose molecules and therefore exposes them to the acid.

Meanwhile, Schüth and co-workers demonstrated that styrene-divinylbenzene resins functionalized with sulfonic groups ($-SO_3H$) are powerful catalysts for breaking down cellulose dissolved in the ionic liquid [C₄mim]Cl into monosaccharides and disaccharides.[11] The team reported that the solid acid resin–ionic liquid process catalyses the hydrolysis of not only purified cellulose extracted from wood but also the cellulose present in untreated wood chips.

In a separate development, ionic liquids have also been shown to be good solvents for the catalytic conversion of sugars such as glucose and fructose into 5-hydroxymethylfurfural (HMF) (**12.2**).[12] HMF could potentially be used in the chemical industry to manufacture polymers and a range of fine chemicals. Good yields of the compound were obtained from glucose using $CrCl_2$ as a catalyst and [C₂mim]Cl as a solvent. $CrCl_2$ and other metal halides such as $FeCl_3$ and $CuCl_2$ also catalysed the conversion of fructose into HMF.

12.2

The use of ionic liquids for the extraction and conversion of naturally occurring compounds is not confined to carbohydrates. In 2000, for example, Lye and co-workers reported that ionic liquids could be used as replacements for organic solvents in multiphase bioprocess operations.[13] The team showed that [C$_4$mim][PF$_6$] can be used to extract erythromycin-A from aqueous solutions. Erythromycin-A is a naturally occurring polyketide antibiotic.

In the same paper, the team reported that ionic liquids can not only be used for liquid–liquid extraction but also for two-phase biotransformation processes, such as enzyme-catalysed transformations.

12.4   ENZYME CATALYSIS

Enzymes are nature's catalysts. They are the proteins that catalyse most chemical reactions in the cells of living systems. In biotechnological applications, enzymatic catalysis is carried out either using fermentation, where a desired product is extracted from a fermentation broth, or by the biotransformation of a starting material into the desired product. Whole cell biocatalysts, such as yeast and various types of bacteria, are used for both fermentations and biotransformations. However, most biotransformations are carried out using isolated enzymes such as lipases, proteases or esterases.

The enzymes used in biotransformations function naturally in aqueous environments. Enzymes, such as lipases, proteases or esterases, that catalyse hydrolysis are known collectively as hydrolases. However, the range of biotransformations catalysed by enzymes in aqueous media is limited owing to the poor solubility of many substrates in water, hydrolysis of the products and other issues such as microbial contamination of the media.

Many hydrolases are stable in anhydrous organic solvents and can therefore be used to catalyse non-hydrolytic reactions in these media. The use of hydrophobic organic solvents in place of aqueous media for biotransformations has several advantages: hydrolytic side reactions are suppressed and hydrophobic substrates and products such as aliphatic, aromatic and heterocyclic compounds dissolve in the solvents. Non-aqueous solvents are also unlikely to suffer from microbial contamination.

The main disadvantage of organic solvents is that they are environmentally unfriendly because of their volatility, toxicity and flammability.

Ionic liquids are potentially environmentally-friendly replacements for "non-green" volatile organic solvents (Chapter 6). In certain cases, enzymes exhibit comparable if not better activities and enhanced enantioselectivities in ionic liquids than in organic solvents. However, the behaviour of enzymes in ionic liquids is not well understood and, as yet, there are no well-defined rules for designing or selecting ionic liquids for specific biotransformations.

Even so, it is possible to make a few generalizations. First, ionic liquids can be designed or selected to dissolve highly polar substrates, such as carbohydrates, that are only sparingly soluble in organic solvents. Ionic liquids also have the advantage over organic solvents in that they are able to stabilize enzymes against thermal inactivation and prevent their thermal degradation.[14] Enzymes therefore remain active in ionic liquids at higher temperatures.

The lack of solubility of proteins in general and enzymes in particular in pure ionic liquids is an important consideration, as it is with organic solvents. In 2001, Kimizuka and Nakashima reported that proteins such as myoglobin, haemoglobin and catalase are insoluble in ionic liquids.[2] They observed, however, that glucose oxidase, a glycosylated protein, is soluble in ionic liquids with ether-containing cations. A few other examples of proteins that dissolve in ionic liquids have been reported but the solubilities tend to be low.

When used for biotransformations in non-aqueous media such as organic solvents, enzymes are typically immobilized on an inorganic support or used in their native form as suspensions. They can therefore be considered as heterogeneous catalysts in both cases.

In general, isolated enzymes form suspensions in ionic liquids and therefore act as heterogeneous catalysts. However, their activities are usually modest. The activity can, in many cases, be enhanced by adding a small amount of water to the ionic liquid or by immobilization of the enzyme on a support. The activity of an enzyme in an ionic liquid may also vary substantially, depending on the ionic liquid and its purity as well as on the method of preparing the enzyme.

In 2006, Goto and co-workers showed that homogeneous enzymatic reactions could be carried out in ionic liquids by modifying the enzyme with a form of poly(ethylene glycol) that has the shape of a comb.[15] They showed that when the protease subtilisin *Carlsberg* is covalently attached to the comb-shaped polymer, it dissolves in three different ionic liquids, all of which contain the anion $[NTf_2]^-$. The team used the transesterification of *N*-acetyl-L-phenylalanine ethyl ester with 1-butanol as a model reaction. In transesterifications, an ester reacts with an alcohol to form a new ester and the alcohol from the original ester.

The activity of the modified enzyme used by Goto's group was much higher in [C$_2$mim][NTf$_2$] than in conventional organic solvents such as toluene. The researchers observed no activity for the native (un-modified) enzyme in the ionic liquids and also no activity for the modified enzyme when hydrophilic organic solvents such as tetrahydrofuran or dimethyl sulfoxide were used. They pointed out that [C$_2$mim][NTf$_2$] is highly polar and hydrophobic. Polar hydrophilic organic solvents such as dimethyl sulfoxide strip essential water from the enzyme and therefore inhibit its activity.

In a recent development, Goto's team demonstrated that the insolubility of enzymes in ionic liquids can also be overcome by using water-in-ionic liquid microemulsions as media for biotransformations.[16] The microemulsions consist of droplets of water in the hydrophobic ionic liquid, [C$_8$mim][NTf$_2$]. The droplets are stabilized by the anionic surfactant sodium bis(2-ethyl-1-hexyl)sulfosuccinate, also known as AOT.

The team showed that various enzymes and proteins dissolve in the water-in-ionic liquid microemulsions but not in the ionic liquid alone or when it is saturated with water. In addition, the researchers showed that the catalytic activity of the lipase-catalysed hydrolysis of *p*-nitrophenyl butyrate in the microemulsions was higher than in microemulsions using an organic solvent instead of the ionic liquid. They observed that the microemulsions could have advantages over microemulsions prepared in conventional organic solvents when used as reaction media for biotransformations of polar or hydrophilic substrates such as amino acids and carbohydrates that are poorly soluble in organic solvents.

12.5 EARLY STUDIES OF IONIC LIQUID–ENZYME SYSTEMS

The study of ionic liquid–enzyme systems dates back to 1984 when an investigation was carried out into the influence of the "the fused salt" ethylammonium nitrate, [C$_2$H$_5$NH$_3$][NO$_3$], on the activity and stability of the enzyme alkaline phosphatase.[17] The aim of the study was to "shed light on the role of solvent in determining protein structure." The investigation found that the enzyme was less active in aqueous solutions of the ionic liquid than in water. The activity decreased as the concentration of the ionic liquid in water increased.

In 1999, it was shown that the same "liquid organic salt" can be used as a precipitating agent for the crystallization of lysozyme, an enzyme that is found in saliva, mucus and hen egg whites.[18]

Enzymes have three-dimensional structures consisting of long folded chains of amino acids. The catalytic activity of enzymes depends on several factors, including the way the chains fold. When the chains become unfolded, by heating for example, the enzymes become denatured, *i.e.* inactive. The activity of some enzymes can be restored by use of an additive that promotes refolding. The refolding process, however, can be inhibited by the formation of aggregates of the inactive enzymes. In 2001, $[C_2H_5NH_3][NO_3]$ was shown to act as a refolding additive that prevents the aggregation of denatured hen egg white lysozyme.[19] The ionic liquid allowed the enzyme to regain 75% of its activity.

The first example of enzyme catalysis using an ionic liquid was described in by Lye and co-workers in a paper, published in 2000, focusing on the use of ionic liquids in multiphase bioprocess operations.[13] The process used the whole cell biocatalyst *Rhodococcus* R312, which is a strain of the *Rhodococcus* bacterium. The bacterium contains the nitrile hydratase enzyme, which facilitates the transformation of nitriles into amides.

The biocatalyst was used in a two-phase process to catalyse the transformation of 1,3-dicyanobenzene into 3-cyanobenzamide and 3-cyanobenzoic acid. $[C_4mim][PF_6]$ was used as a reservoir for the substrate, 1,3-dicyanobenzene. The biocatalyst was present in a separate buffered aqueous phase, not in the ionic liquid phase. The work showed that the activity of the biocatalyst in the water–ionic liquid biphasic system was almost an order of magnitude greater than in a water–toluene biphasic system.

Since then, there have been a few other studies of whole cell-ionic liquid systems. In 2001, for example, baker's yeast was shown to reduce several ketones to the corresponding alcohols in a 10 : 1 mixture of $[C_4mim][PF_6]$ and water.[20] The yeast was immobilized by encapsulation in calcium alginate beads. The yield of the product varied, depending on the ketone. In particular, it was not possible to reduce two aromatic ketones: 4-bromoacetophenone and 4-methoxyacetophenone.

In the same year that Lye's group described their work on whole cell biocatalysis with an ionic liquid, Erbeldinger and co-workers published the first report of enzymatic catalysis in an ionic liquid using an isolated enzyme.[21] The team used the protease thermolysin for the synthesis of the dipeptide Z-aspartame (**12.3**), where Z is the carbobenzyloxy protecting group, from two amino acid derivatives: carbobenzyloxy-L-aspartate (**12.4**) and L-phenylalanine methyl ester (**12.5**) (Scheme 12.1). The dipeptide is a precursor to the artificial sweetener aspartame. The reaction was carried out in $[C_4mim][PF_6]$ containing 5% water by volume. The researchers found that the rate of the enzymatic synthesis in

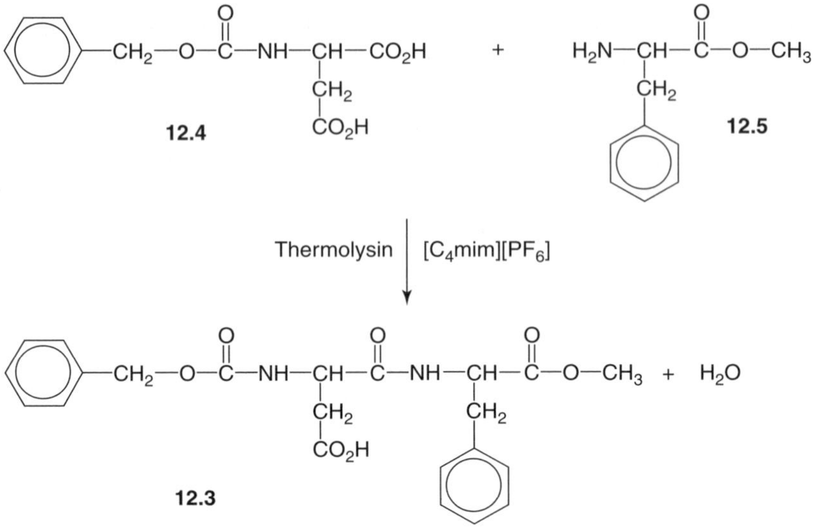

Scheme 12.1 Thermolysin catalyses the synthesis of (Z)-aspartame in an ionic liquid.

the ionic liquid was comparable to that in the organic solvent ethyl acetate. They also observed "excellent" stability of the enzyme in the ionic liquid.

12.6 LIPASES

The first example of enzymatic catalysis in an ionic liquid in the absence of added water was also reported in 2000.[22] In this paper, Sheldon and co-workers described the first example of lipase-catalysed reactions in ionic liquids. Lipases are enzymes that catalyse the cleavage of fats and oils in nature. In industry, they are widely used to catalyse esterifications and transesterifications. They are employed, for example, in the synthesis of pharmaceutical and agrochemical intermediates and in the manufacture of flavours, fragrances and health care products.

Sheldon's team studied the catalysis of several types of chemical reaction using *Candida antarctica* lipase B. The researchers reported that the use of the ionic liquids [C$_4$mim][BF$_4$] and [C$_4$mim][PF$_6$] as solvents for these reactions achieved results comparable to those of organic solvents.

They carried out, for example, lipase-catalysed transesterifications in these ionic liquids. In addition, they conducted the lipase-catalysed reaction of octanoic acid with ammonia in [C$_4$mim][BF$_4$]. The ammoniolysis resulted in a quantitative yield of octanamide.

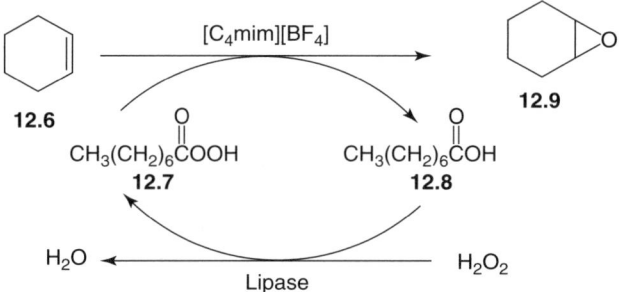

Scheme 12.2 Epoxidation of cyclohexene in an ionic liquid.

The team also investigated the epoxidation of cyclohexene (**12.6**) by peroctanoic acid (**12.7**) in [C$_4$mim][BF$_4$]. The acid was generated *in situ* by the lipase-catalysed perhydrolysis of octanoic acid (**12.8**) with aqueous hydrogen peroxide (Scheme 12.2). The yield of epoxide **12.9** was 83%. In general, the reaction rates were comparable with or better than those observed in conventional organic solvents.

Since the original work by Sheldon's team in 2000, numerous examples of lipase-catalysed reactions in ionic liquids have been reported. In 2008, for example, Klembt and co-authors listed 27 reaction systems that employed lipases in ionic liquids.[23]

In general, lipases maintain their activity in anhydrous ionic liquids and, in many cases, their activities and selectivities are comparable to and sometimes better than in conventional organic solvents.

12.7 PROTEASES, ESTERASES AND OTHER ENZYMES

Biotransformations in ionic liquids using proteases have also been extensively studied over recent years. For example, the activity of the protease α-chymotrypsin in ionic liquids has also been studied by several research teams. This enzyme, which is found in intestines, helps us to digest food by breaking the peptide linkages in proteins.

In 2001, Laszlo and Compton reported an investigation into the α-chymotrypsin catalysis of the transesterification of *N*-acetyl-L-phenylalanine ethyl ester (**12.10**) with propan-1-ol in two water-immiscible imidazolium ionic liquids: [C$_4$mim][PF$_6$] and the less polar [C$_8$mim][PF$_6$].[24] The products of the reaction are the corresponding propyl ester **12.11** and ethanol (Scheme 12.3).

Laszlo and Compton observed no activity of the protease at very low water concentrations. However, when 0.25% water, volume/volume,

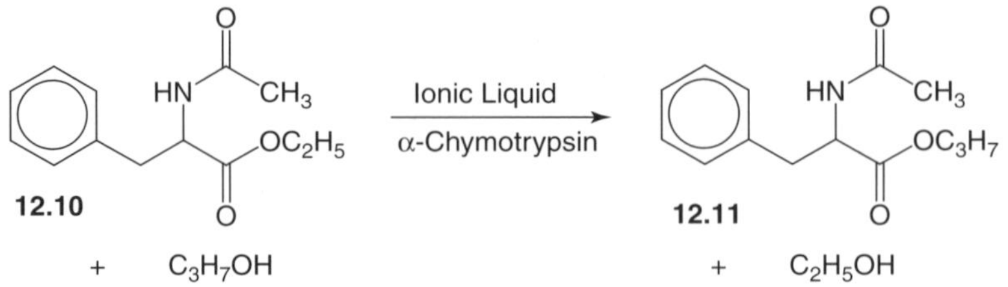

Scheme 12.3   Enzyme catalyses transesterification in an ionic liquid.

was added to the ionic liquids, transesterification occurred at moderate rates. The two researchers observed that neat ionic liquids remove essential water that is bound to the enzyme. A small amount of water in the solvent is therefore desirable to maintain enzyme activity. Better conversions were achieved using the less polar [C$_8$mim][PF$_6$] than with [C$_4$mim][PF$_6$], the results being comparable to those achieved with a non-polar solvent such as hexane. Non-polar solvents are less likely to strip water from the enzyme. However, added water was not required if supercritical CO$_2$, which is non-polar, was used as a co-solvent with the ionic liquid.

There have been relatively few studies of biotransformations in ionic liquids using enzymes other than lipases or proteases. This is because other types of enzyme show little activity in non-aqueous media.

In one of the investigations of other types of enzymes, published in 2003, Persson and Bornscheuer demonstrated that two esterases catalyse the transesterification of 1-phenylethanol with vinyl acetate in three imidazolium ionic liquids.[25] The esterases were from two species of bacteria: *Bacillus subtilis* and *B. stearothermophilus*. The ionic liquids were [C$_4$mim][PF$_6$], [C$_4$mim][BF$_4$] and [C$_4$mim][NTf$_2$].

The two researchers observed no transesterification activity in the ionic liquids when a lyophilized, *i.e.* freeze dried, powder of the esterases was used. However, when the esterases were immobilized on Celite (diatomaceous earth), the activity was comparable to that obtained in the organic solvent methyl *tert*-butyl ether (MTBE) but less than that in *n*-hexane. The enantioselectivity did not depend on which organic or ionic liquid solvent was used. The stability of the esterase from *Bacillus stearothermophilus* proved far greater in [C$_4$mim][PF$_6$] and [C$_4$mim][BF$_4$].

The activity of some other classes of enzymes has also been examined in ionic liquids. For example, Kragl and co-workers studied the use of the peptide amidase from the bacterium *Stenotrophomonas maltophilia*.[26] The enzyme, a hydrolase, catalyses the hydrolysis of the terminal amide groups in peptides. It can also catalyse the reverse reaction: the conversion of the terminal carboxyl group (-COOH) of a peptide into an amide (-CONH$_2$). The process is known as C-terminal amidation.

The team investigated the amidation of the dipeptide H$_2$N-Ala-Phe-COOH to H$_2$N-Ala-Phe-CONH$_2$ in several ionic liquids using NH$_4$HCO$_3$ as the source of ammonium and lyophilized amidase as the catalyst. The researchers also examined the influence of ionic liquid and water concentration on the amidation activity using various ionic liquids. The maximum yield of the amide was obtained using the ionic liquid [C$_4$mim][CH$_3$SO$_4$] containing virtually no water. The reaction rates for the amidations were comparable to those observed for conventional organic solvents.

12.8   ENANTIOSELECTIVITY AND REGIOSELECTIVITY

Over half of the 27 the lipase–ionic liquid systems listed by Klembt and co-authors in their 2008 article[23] focused on kinetic resolution, a process that relies on the different reaction rates of the two enantiomers in a racemic mixture. Enantiomers are molecules that are mirror-images of one another. If the enantiomers are in dynamic equilibrium with one another during a reaction, the slower reacting enantiomer converts into the faster one as the reaction proceeds. This process is known as dynamic kinetic resolution.

One of the earliest papers on the topic described an investigation into the dynamic kinetic resolution of racemic 1-phenylethanol (**12.12**) by transesterification with vinyl acetate (**12.13**) in a range of imidazolium ionic liquids, using various lipases as biocatalysts (Scheme 12.4).[27] The

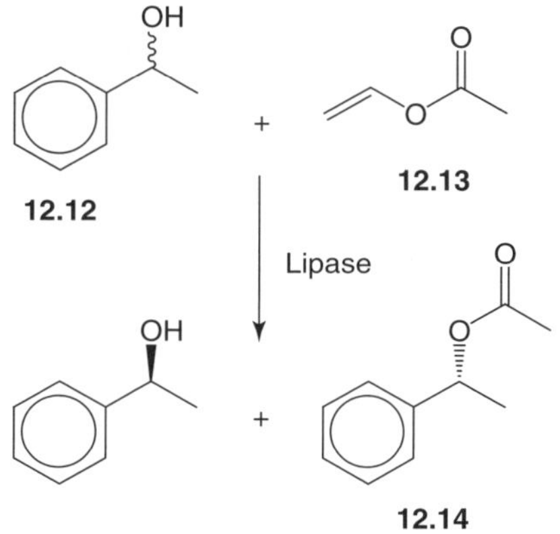

Scheme 12.4   Dynamic kinetic resolution of 1-phenylethanol by transesterification.

percentage conversions of the racemic mixture and enantiomeric selectivity for the formation of the (*R*)-acetate **12.14** were compared with
the results for the biocatalysed reaction in the organic solvent MTBE.
Some of the lipases showed little or no catalytic activity while others
exhibited good activities. In some cases the enantioselectivities were
better than those for MTBE. The researchers also showed that the enzyme suspended in the ionic liquid could be reused three times with less
than 10% loss of activity per cycle.

In the same year, Park and Kazlauskas reported on the *Pseudomonas
cepacia* lipase-catalysed acetylation of 1-phenylethanol with vinyl acetate.[28] They discovered that the reaction was as fast and as enantioselective
in purified imidazolium and pyridinium ionic liquids as in non-polar
solvents such as toluene. With impure ionic liquids, however, the reactions were slower than in toluene or did not occur at all.

The paper also investigated enzyme regioselectivity in ionic liquids.
Regioisomers are molecules with substituents located at different positions. Park and Kazlauskas showed that the acetylation of glucose with
vinyl acetate catalysed by *Candida antarctica* lipase B was more regioselective in imidazolium and pyridinium ionic liquids with $[BF_4]^-$ anions
than in organic solvents such as tetrahydrofuran and acetone. In the
organic solvents the lipase-catalysed reaction resulted in a mixture of the
6-*O*-acetylglucose (**12.15**) and 3,6-*O*-diacetylglucose (**12.16**), whereas in
the ionic liquids the product was predominantly 6-*O*-acetylglucose. The
two researchers attributed the increase in selectivity for the 6-*O*-acet-
ylglucose regioisomer to the increased solubility of glucose in the ionic
liquid compared with organic solvents.

12.15   **12.16**

The influence of ionic liquids on the regioselectivity of lipase-catalysed
reactions was also demonstrated by Salunkhe and co-workers.[29] They
studied the *Pseudomonas cepacia* lipase-catalysed regioselective deace-
tylation of 3,4,6-tri-*O*-acetyl-D-glucal (**12.17**) by hydrolysis and alcoholysis in [C₄mim][PF₆], [C₄mim][BF₄] and the organic solvent
tetrahydrofuran (Scheme 12.5). Hydrolysis was performed using a
phosphate buffer whereas the alcoholysis used decanol instead of the
buffer. The hydrophilic ionic liquid [C₄mim][BF₄] proved not to be a

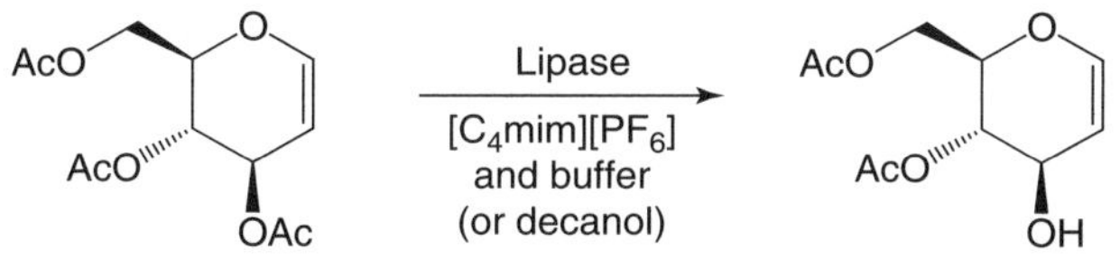

Scheme 12.5 Lipase-mediated regioselective hydrolysis (alcoholysis).

good solvent for the biotransformations whereas the hydrophobic ionic liquid [C$_4$mim][PF$_6$] exhibited "remarkable" regioselectivity for the desired product 4,6-di-*O*-acetyl-D-glucal (**12.18**).

More recently, Salunkhe's group designed and synthesized a task-specific ionic liquid for the enzyme-catalysed kinetic resolution of ibuprofen (**12.19**).[30] Ibuprofen is a non-steroidal anti-inflammatory drug. The compound has a chiral centre in its propionic acid side chain. However, only the (*S*)-enantiomer acts as an anti-inflammatory agent.

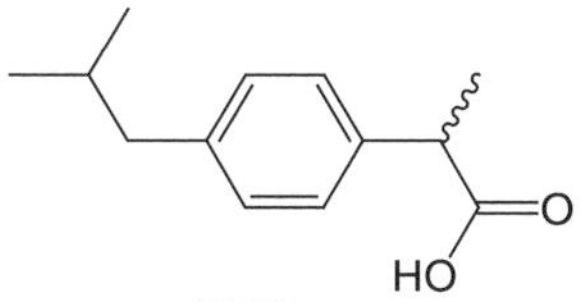

Salunkhe's team appended a hydroxyl group to the butyl side chain of the cation in the ionic liquid [C$_4$mim][PF$_6$]. Racemic ibuprofen was then covalently attached to the ionic liquid. *Candida antarctica* lipase-catalysed hydrolysis of the ionic liquid anchored ibuprofen led to good yields of and high enantioselectivity for the (*S*)-enantiomer.

12.9 BIPHASIC SYSTEMS AND SEPARATIONS

Investigations of the use of ionic liquids for biotransformations have generally focused on their use as single phases, either in their pure form or in combination with another solvent such as water. However, the first reported example of enzyme catalysis using an ionic liquid employed a biphasic system (Section 12.5). The ionic liquid acted as a reservoir for the substrate and a buffered aqueous phase as a medium for a whole cell catalyst.

In 2004, Kragl and co-workers described the enzyme-catalysed enantioselective reduction of a ketone in a biphasic system consisting of the ionic liquid [C$_4$mim][NTf$_2$] and a buffer.[31] They showed that the alcohol dehydrogenase from the lactic acid bacteria *Lactobacillus brevis* catalyses the reduction of 2-octanone to (*R*)-2-octanol (**12.20**). The reduction proceeded more rapidly in the ionic liquid than in the organic solvent MTBE.

12.20

The ionic liquid phase acted as a reservoir for the poorly water-soluble substrate, 2-octanone. The aqueous phase contained the enzyme, its cofactor NADPH (the reduced form of nicotinamide adenine dinucleotide phosphate), and 2-propanol, which was used as a co-substrate to regenerate the cofactor. During the process, 2-propanol is oxidized to acetone, which is a known inhibitor for alcohol dehydrogenase. However, because of favourable partition coefficients, the acetone was removed to the ionic liquid phase, leaving little in the aqueous phase containing the enzyme and cofactor.

One of the simplest ways to recover a reaction product from a homogeneous solution is to use a solvent that is immiscible with the solution. A biphasic system is formed in which the product preferentially dissolves in the extracting solvent. However, extracting solvents are generally volatile organic compounds and therefore "non-green." One way of overcoming this problem is to use supercritical CO$_2$ as an environmentally benign extracting solvent.

In 2002, Reetz and co-workers reported that supercritical CO$_2$ can be used as a vehicle not only to carry reactants into an ionic liquid containing an enzyme but also to extract the biotransformation products from the ionic liquid.[32] The team used *Candida antarctica* lipase B to catalyse the acylation of octan-1-ol by vinyl acetate in [C$_4$mim][NTf$_2$] as a model reaction. The substrates, octan-1-ol and vinyl acetate, were introduced into a stream of supercritical CO$_2$. The mixture was then passed through a continuous flow reactor containing a mixture of the ionic liquid and the enzyme. The CO$_2$ extracted the products from the reaction mixture and, after passing out of the reactor, was depressurized. The products were collected in a cold trap. Yields of octyl acetate in excess of 90% were obtained. Furthermore, the team showed that the enzyme–ionic liquid mixture can be recycled many times without loss of activity.

In 2004, Iborra and co-workers reported the use of a continuous flow enzymatic reactor apparatus for the *Candida antarctica* lipase B-catalysed kinetic resolution of racemic 1-phenylethanol by transesterification with vinyl propionate in ionic liquid–supercritical CO_2 biphasic systems.[33] The researchers tested five ionic liquids with $[NTf_2]^-$ anions and various quaternary ammonium cations with functional side chains. The ionic liquids proved not only to exhibit very high enantioselectivity for the (S)-1-phenylethyl propionate but also to be "excellent" agents for stabilizing the enzyme.

Supported ionic liquid membranes (Section 5.19) have also been shown to promote enantioselectivity. In 2003, for example, Goto and co-workers used SILMs to facilitate the enantioselective production of (S)-ibuprofen.[34] The SILMs separated an aqueous feed phase from an aqueous receiving phase. The membranes were prepared by immersing a hydrophobic poly(propylene) film in an imidazolium ionic liquid such as $[C_4mim][NTf_2]$. The feed phase contained racemic ibuprofen (**12.19**) and ethanol as substrates and lipase from *Candida rugosa*. The enzyme catalyses the esterification of (S)-ibuprofen but not (R)-ibuprofen. The (S)-ibuprofen ester is selectively transported through the SILM into the receiving phase, which contains the enzyme porcine pancreas lipase. This enzyme catalyses the hydrolysis of the (S)-ibuprofen ester to form (S)-ibuprofen and ethanol.

REFERENCES

1. S. Park and R. J. Kazlauskas, *J. Org. Chem.*, 2001, **66**, 8395.
2. N. Kimizuka and T. Nakashima, *Langmuir*, 2001, **17**, 6759.
3. S. A. Forsyth, D. R. MacFarlane, R. J. Thomson and M. von Itzstein, *Chem. Commun.*, 2002, 714.
4. R. P. Swatlowski, S. K. Spear, J. D. Holbrey and R. D. Rogers, *J. Am. Chem. Soc.*, 2002, **124**, 4974.
5. N. Sun, R. P. Swatloski, M. L. Maxim, M. Rahman, A. G. Harland, A. Haque, S. K. Spear, D. T. Daly and R. D. Rogers, *J. Mater. Chem.*, 2008, **18**, 283.
6. D. A. Fort, R. P. Swatlowski, P. Moyna, R. D. Rogers and G. Moyna, *Chem. Commun.*, 2006, 714.
7. D. A. Fort, R. C. Remsing, R. P. Swatloski, P. Moyna, G. Moyna and R. D. Rogers, *Green Chem.*, 2007, **9**, 63.
8. A. G. Fadeev and M. M. Meagher, *Chem. Commun.*, 2001, 295.
9. M. Matsumoto, K. Mochiduki, K. Fukunishi and K. Kondo, *Sep. Purif. Technol.*, 2004, **40**, 97.
10. C. Li, Q. Wang and Z. K. Zhao, *Green Chem.*, 2008, **10**, 177.

11. R. Rinaldi, R. Palkovits and F. Schüth, *Angew. Chem. Int. Ed.*, 2008, **47**, 8047.
12. H. Zhao, J. E. Holladay, H. Brown and Z. C. Zhang, *Science*, 2007, **316**, 1597.
13. S. G. Cull, J. D. Holbrey, V. Vargas-Mora, K. R. Seddon and G. J. Lye, *Biotechnol. Bioeng.*, 2000, **69**, 227.
14. G. A. Baker, S. N. Baker, T. M. McCleskey and J. H. Werner, in *Ionic Liquids as Green Solvents: Progress and Prospects*, ed. R. D. Rogers and K. R. Seddon, ACS Symposium Series 856, American Chemical Society, Washington DC, 2003, p. 212.
15. K. Nakashima, T. Maruyama, N. Kamiya and M. Goto, *Org. Biomol. Chem.*, 2006, **4**, 3462.
16. M. Moniruzzaman, N. Kamiya, K. Nakashima and M. Goto, *Green Chem.*, 2008, **10**, 497.
17. D. K. Magnuson, J. W. Bodley and D. F. Evans, *J. Solution Chem.*, 1984, **13**, 583.
18. J. A. Garlitz, C. A. Summers, R. A. Flowers II and G. E. O. Borgstahl, *Acta Crystallogr. Sect. D*, 1999, **55**, 2037.
19. C. A. Summers and R. A. Flowers II, *Protein Sci.*, 2000, **9**, 2001.
20. J. Howarth, P. James and J. Dai, *Tetrahedron Lett.*, 2001, **42**, 7517.
21. M. Erbeldinger, A. J. Mesiano and A. J. Russell, *Biotechnol. Prog.*, 2000, **16**, 1129.
22. R. M. Lau, F. van Rantwijk, K. R. Seddon and R. A. Sheldon, *Org. Lett.*, 2000, **2**, 4189.
23. S. Klembt, S. Dreyer, M. Eckstein and U. Kragl, in *Ionic Liquids in Synthesis*, 2nd edn, ed. P. Wasserscheid and T. Welton, Wiley-VCH, Weinheim, 2008, **vol. 2**, p. 641.
24. J. A. Laszlo and D. L. Compton, *Biotechnol. Bioeng.*, 2001, **75**, 181.
25. M. Persson and U. T. Bornscheuer, *J. Mol. Catal B: Enzym.*, 2003, **22**, 21.
26. N. Kaftzik, S. Neumann, M.-R. Kula and U. Kragl, in *Ionic Liquids as Green Solvents Progress and Prospects*, ed. R. D. Rogers and K. R. Seddon, ACS Symposium Series 856, American Chemical Society, Washington DC, 2003, p. 206.
27. S. H. Schöfer, N. Kaftzik, P. Wasserscheid and U. Kragl, *Chem. Commun.*, 2001, 425.
28. S. Park and R. J. Kazlauskas, *J. Org. Chem.*, 2001, **66**, 8395.
29. S. J. Nara, S. S. Mohile, J. R. Harjani, P. U. Naik and M. M. Salunkhe, *J. Mol. Catal. B: Enzym.*, 2004, **28**, 39.
30. P. U. Naik, S. J. Nara, J. R. Harjani and M. M. Salunkhe, *J. Mol. Catal. B: Enzym.*, 2007, **44**, 93.

31. M. Eckstein, M. V. Filho, A. Liese and U. Kragl, *Chem. Commun.*, 2004, 1084.
32. M. T. Reetz, W. Wiesenhöfer, G. Franciò and W. Leitner, *Chem. Commun*, 2002, 992.
33. P. Lozano, T. de Diego, S. Gmouh, M. Vaultier and J. L. Iborra, *Biotechnol. Prog.*, 2004, **20**, 661.
34. E. Miyako, T. Maruyama, N. Kamiya and M. Goto, *Chem. Commun.*, 2003, 2926.

Analysis

13.1 INTRODUCTION

Various conventional instrumental methods have been employed for the separation, identification and characterization of ionic liquids, for the assessment of their purity, and for the determination of the type and nature of impurities present (Chapter 4). Such methods include titration, thermogravimetric analysis, mass spectrometry, neutron scattering, capillary electrophoresis and various forms of voltammetry, chromatography and spectroscopy. For example, ion chromatography, electrophoresis and voltammetry have been used to determine chloride impurities in both water-miscible and water-immiscible ionic liquids.

This chapter focuses principally on the use of ionic liquids as solvents and electrolytes for various analytical techniques, rather than on the analysis of ionic liquids themselves.

13.2 CHROMATOGRAPHY

The thermal stability, viscosity, polarity, solubility and partitioning properties of ionic liquids combined with their negligible vapour pressures make the liquids attractive for use in gas chromatography (GC) and liquid chromatography (LC). In these forms of chromatography, a mobile phase, the gas or liquid, respectively, which carries the substance to be analysed (the analyte), is passed through a column containing a solid stationary phase that retains and separates the components of the analyte.

An Introduction to Ionic Liquids
By Michael Freemantle
© Michael Freemantle 2010
Published by the Royal Society of Chemistry, www.rsc.org

Various studies have investigated the use of ionic liquids for coated stationary phases for GC. The ionic liquids are typically coated on a solid column support such as fused-silica capillary tubing. The use of ionic liquids for LC, either as stationary phases or for the modification of mobile phases, has been studied to a lesser extent.

Some of the earliest work, carried out in the early 1980s, focused on the use of the ionic liquids ethylammonium nitrate and ethylpyridinium bromide as stationary phases for GC. The first report, which appeared in 1982, showed that ethylammonium nitrate was suitable for use as a GC stationary phase for the separation of a mixture of alcohols or a mixture of monofunctional benzene derivatives.[1] The use of this ionic liquid as a stationary phase is limited however. Proton-acceptors such as amines and proton donors such as phenol cannot be eluted from the column. Alkanes and alkenes, on the other hand, are not retained by the stationary phase.

In the late 1990s, attention turned to the use of imidazolium ionic liquids, such as $[C_4mim][PF_6]$ and $[C_4mim]Cl$, coated on silica as GC stationary phases to separate volatile organic compounds.[2] These ionic liquids were shown to exhibit dual stationary phase behaviour. They act as non-polar stationary phases for separating non-polar compounds and polar compounds that are poor proton donors or acceptors. On the other hand, they act as polar stationary phases for separating polar compounds with proton donor or acceptor groups.

Ionic liquids have also been tested as GC stationary phase solvents for chiral selectors, *i.e.* compounds that separate chiral compounds. For example, Berthod and co-workers prepared chiral separation stationary phases for capillary columns by dissolving methylated cyclodextrins, the chiral selectors, in $[C_4mim]Cl$, an achiral room-temperature ionic liquid.[3] The ionic liquid–cyclodextrin mixture was coated onto the walls of the capillary columns. However, the ability of the mixture to resolve enantiomers proved not to be as good as commercial columns with the same chiral selectors.

The use of ionic liquids for high-performance liquid chromatography (HPLC) has also been the subject of several investigations. For example, Sun and Stalcup showed that $[C_4mim]Br$, covalently immobilized by an alkyl tether on a silica substrate, is effective as a HPLC stationary phase when used with binary acetonitrile–water or methanol–water mobile phases.[4] The two chemists concluded from their work that "this new stationary phase shows promise for novel separations."

The use of 1-alkyl-3-methylimidazolium and 1-butylpyridinium ionic liquids as additives to HPLC mobile phases has also been examined. Zhang and co-workers demonstrated that a HPLC mobile phase consisting of water with millimolar concentrations of these ionic liquid additives is effective at separating catecholamines such as adrenaline and dopamine.[5]

Whereas LC traditionally uses a solid stationary phase, counter-current chromatography employs liquid stationary and liquid mobile phases. The relatively high viscosities of ionic liquids severely restrict the use of pure ionic liquids as stationary phases in this type of chromatography. However, the addition of an organic solvent such as acetonitrile reduces the viscosity.

Berthod and co-workers have tested a biphasic $[C_4mim][PF_6]$–acetonitrile–water system for counter-current chromatography.[6] The denser ionic liquid-rich phase and upper aqueous phase were used, respectively, as the stationary and mobile phases. The authors observed that the separation capability of the system was "not fully exploited" in counter-current chromatography, adding that "further studies are needed."

13.3 CAPILLARY ELECTROPHORESIS

Ionic liquids not only exhibit ionic conductivity but can also dissolve a wide range of substances. These properties have led to several studies of their use as electrolytes for capillary electrophoresis (CE).

CE is an analytical separation technique that relies on the migration of charged analytes under the influence of an externally applied electric field through a capillary tube that is typically made of fused silica. As they migrate, the analytes separate owing to differences in their electrophoretic mobilities. These depend on their size-to-charge ratios. The separated analytes are detected by UV/Vis spectroscopy or some other technique.

CE requires an electrolytic solvent, normally an aqueous buffer solution, for the sample under investigation. The buffer solution is sometimes known as the "run buffer" or "running electrolyte" because it flows through the capillary when a potential is applied. This bulk flow is called electroosmotic flow.

Aqueous buffers are not suitable as CE solvents for analytes that are poorly soluble in water. In these cases, electrolyte solutions prepared from organic solvents such as acetonitrile, propylene carbonate or methanol are used. The technique is then known as non-aqueous capillary electrophoresis.

In both aqueous and non-aqueous CE, a background electrolyte such as ammonium acetate is commonly added to enhance the separation of the analytes. It is in this role as background electrolytes for both aqueous and non-aqueous forms of the technique that ionic liquids have been most widely studied.

Various ionic liquids have been used as background CE electrolytes to promote the separation of neutral analytes such as phenolic compounds. For example, in 2001, Stalcup and co-workers reported the use of aqueous solutions of $[C_4mim][BF_4]$ and other water-miscible imidazolium ionic liquids as running electrolytes for the CE of neutral compounds.[7] They used the technique to separate polyphenolic compounds, such as catechin, found in grape seed extracts. The authors suggested that the separation mechanism involves association between the imidazolium cations and the polyphenol molecules.

In related work, Kuldvee and colleagues employed imidazolium ionic liquids as background electrolytes for non-aqueous CE to examine the separation of neutral methyl- and hydroxyl-substituted phenolic compounds.[8] They used acetonitrile and propylene carbonate as the organic solvents for the running electrolyte. The researchers concluded from their work that, in this case, the separation of the neutral phenolic compounds arises from the formation of a negative complex between the molecules and the background electrolyte anions.

In general, the miscibility of ionic liquids, notably imidazolium ionic liquids, with organic solvents allows the electrophoretic mobilities and separation of analytes to be optimized. For example, Vaher and Koel showed that the selection of a $[C_4mim]^+$ ionic liquid with the heptafluorobutanoate anion, $[CF_3(CF_2)_2CO_2]^-$, as a background electrolyte and methanol as a solvent enables small metal cations such as Na^+ and Mg^{2+} to be separated by CE.[9]

Warner and co-workers have used ionic liquids in a capillary electrophoretic technique known as micellar electrokinetic chromatography for the separation of achiral and chiral analytes.[10] The technique is widely used for separating uncharged analytes in aqueous solutions. Typically, a surfactant such as sodium dodecyl sulfate is added to the buffer solution to form micelles. The anionic sulfate groups of the micelle give rise to electrophoretic mobility in the opposite direction to the electroosmotic flow of the buffer. The micellar phase is considered to be pseudo-stationary as it migrates much slower than the buffer. Analytes dissolved in the micelle migrate with the micelle in one direction when a potential is applied. Analytes that are insoluble in the micelles migrate in the buffer with the electroosmotic flow.

Warner's team showed that the addition of the hydrophilic ionic liquids $[C_2mim][BF_4]$ or $[C_4mim][BF_4]$ as pseudo-stationary phase modifiers to a background electrolyte buffer solution containing an achiral sulfate-containing polymeric surfactant enhanced the separation of achiral mixtures of alkyl aryl ketones and chlorophenols. The same ionic liquids, when used with a chiral polymeric surfactant, also facilitated the resolution of enantiomers of chiral binaphthyl derivatives.

Ionic liquids have also been used as support coatings on the inner walls of silica capillaries used for CE. Silanol (SiOH) groups on the surfaces of these walls generate negative charges. As a consequence, the walls adsorb analyte cations resulting in impaired separation efficiency of the analytes.

One method of overcoming this problem is to reverse the surface charge and the electroosmotic flow of the buffer by covalently coating the inner capillary surface with positively charged ions. Qin and Li have used ionic liquid cations for this purpose.[11] They showed that immobilization of $[C_6mim]^+$ cations on the inner capillary surface reversed the electroosmotic flow, enabling a mixture of metal cations to be separated efficiently.

13.4 SENSORS

Chemical sensors are used extensively for monitoring and controlling chemicals in a wide variety of industrial, medical, environmental, domestic and other applications. For example, the use of chemical sensors to determine ammonia is important in the food industry for detecting food decomposition, in the automotive industry for checking ammonia emissions from vehicles, in the chemical industry for processes where ammonia is a reactant or product and in medical diagnosis. To be effective, such sensing devices need to function rapidly, to be sensitive and selective for a specific analyte, and they also need to be reliable and inexpensive.

Over the past decade, several research groups around the world have been attempting to exploit the solubility, ionic conductivity and other properties of ionic liquids to develop sensor devices for detecting various gases, vapours and ions in solution.

13.5 GAS SENSORS

Much of the research on using ionic liquid-based chemical sensors has focused on the detection of gases.

One such sensor, a quartz crystal microbalance (QCM) device, exploits the varying solubilities of gases in ionic liquids. QCM devices have quartz crystal resonators that change in frequency when their masses change.

The first ionic liquid-based QCM sensor was developed and evaluated by Dai and colleagues.[12] They used the device to sense organic vapours such as acetone, methanol, toluene and chloroform.

The device has a resonator coated with an imidazolium ionic liquid such as $[C_4mim][BF_4]$ or $[C_4mim]NTf_2]$. In the presence of an organic vapour, the frequency of the resonator rapidly changes. The frequency shift depends on the ability of an ionic liquid film to dissolve the vapour. Different ionic liquids give different frequency shifts for a specific vapour and a specific ionic liquid gives different frequency shifts for different vapours. The team attribute the varying frequency shifts to the varying decreases in viscosity and density of the ionic liquids when they absorb the vapours.

After use, the ionic liquid is readily rinsed off and the quartz crystal re-used. In principle, an ionic liquid can be selected or tailored to respond to a specific target organic vapour. Furthermore, an array of such devices with different ionic liquids could be used to selectively monitor a range of target vapours.

Baltus and co-workers used a similar QCM device to determine the solubility of CO_2 in a series of imidazolium ionic liquids at low pressures.[13] The results showed that the CO_2 solubility increased as the length of the alkyl side chain on the imidazolium ring increased. The solubility was also significantly greater in an ionic liquid with a fluorine-substituted cation than with the corresponding ionic liquid with a non-fluorinated cation. In addition, solubility was found to be lower in ionic liquids with $[PF_6]^-$ anions than with the corresponding liquids with $[NTf_2]^-$ anions.

Commercial gas sensors commonly use potentiometric or amperometric methods to detect a gas. The gas diffuses through a gas-porous hydrophobic membrane into an electrolyte solution where it is oxidized or reduced at an electrode, thereby generating a cell potential and an external current. In electrochemical carbon monoxide sensors, for example, CO is oxidized to CO_2 at one electrode and oxygen is consumed at the other.

The CO device and other electrochemical gas sensors typically employ an aqueous solution of a mineral acid such as sulfuric acid as the electrolyte. As water evaporates, the sensor eventually dries out, limiting its lifetime. Ionic liquids, on the other hand, have negligible vapour pressure and, because of their intrinsic ionic conductivity, they do not require the addition of a supporting electrolyte. Furthermore, the thermal stability of ionic liquids means that ionic liquid-based gas sensors can be used at high temperatures. Their main disadvantage is the low rates of diffusion of the gaseous analyte in the ionic liquids, especially those with higher viscosities.

In 2004, Ohsaka and co-workers described the development of a novel amperometric oxygen gas sensor based on [C$_2$mim][BF$_4$] supported on porous polyethylene membrane-coated electrodes.[14] The current generated by amperometric gas sensors is directly proportional to the gas concentration. The sensors are particularly attractive as they respond rapidly to gases and exhibit high sensitivity over a wide range of gas concentrations. The ionic liquid-based O$_2$ sensor proved to have a wide detection range, high sensitivity and was easily constructed, according to the authors.

Its main drawback, however, was that it only detected O$_2$ in dry gas streams. Water absorbed by [C$_2$mim][BF$_4$], a water-miscible ionic liquid, interfered with the detection. The team subsequently tested the hydrophobic ionic liquid [C$_4$mim][PF$_6$] in place of [C$_2$mim][BF$_4$].[15] Once again the presence of water in the electrode system was shown to interfere with the results, restricting the use of the sensor to dry gas streams.

A group of chemists led by Compton has investigated the use of ionic liquids as electrolytes for sensing a range of gases: O$_2$, CO$_2$, H$_2$, NH$_3$, SO$_2$ and NO$_2$. For example, the group studied the electrochemical oxidation of hydrogen at activated platinum electrodes in ten dialkylimidazolium, tetraalkylammonium and tetralkylphosphonium room-temperature ionic liquids.[16] They concluded that the large number of room-temperature ionic liquids "allows scope for the possible electrochemical detection of hydrogen gas for use in gas sensor technology."

Compton and colleagues also investigated the electrochemical reduction of sulfur dioxide, SO$_2$, an air pollutant that causes acid rain, in various imidazolium ionic liquids, including [C$_4$mim][NO$_3$].[17] They observed that the voltammetry of SO$_2$ in [C$_4$mim][NO$_3$] "loosely resembles that obtained in conventional aprotic solvents" such as dimethyl formamide and dimethyl sulfoxide. The high sensitivity of the [C$_4$mim][NO$_3$] system to SO$_2$ suggests that the ionic liquid may be a viable solvent in gas sensing applications, according to the authors.

In related work, Compton and co-workers also studied the electrochemical reduction of hydrogen sulfide, H$_2$S, on a platinum electrode in five imidazolium room-temperature ionic liquids.[18] H$_2$S is a highly toxic gas that occurs naturally in coal pits and oil wells. Compton's group noted that the solubilities of the gas in ionic liquids are much higher than those reported in conventional molecular solvents. The authors suggested that the liquids could prove to be "very favourable gas sensing media for H$_2$S detection." The group attributed the reduction of H$_2$S to the formation of hydrogen and HS$^-$ anions at the electrode.

13.6 ION-SELECTIVE ELECTRODES

The incorporation of ionic liquids into ion-selective electrodes (ISEs) to enhance their response towards specific ions has been investigated by a few research groups in recent years.

ISEs generally employ a membrane, typically a glass or polymer membrane, that senses a specific ion. For example, the glass pH electrode is an ISE that is sensitive to H^+ ions. When ions enter the membrane from the solution under investigation, an electric potential is generated that is proportional to the concentration of the ion. Many of the ISEs that are used to sense anions have plasticized poly(vinyl chloride) (PVC) membranes. These membranes contain an ionophore that allows the selected ions to pass through the membrane.

In 2005, Soto, Mañez and colleagues reported that plasticized PVC membranes containing [C_4mim][PF_6] and a polyazacycloalkane ionophore (**13.1**) led to a "remarkable enhanced response" towards sulfate anions in aqueous solutions.[19] The paper was the first report of the use of ionic liquids as membrane components for ISEs. The authors predicted that the large number of known ionic liquids and the possibility of incorporating them into the membranes might lead to a new generation of polymer-based ISEs with enhanced responses towards hydrophilic anions.

13.1

The following year, Pletnev and colleagues showed that imidazolium and phosphonium ionic liquids could be incorporated in PVC and poly(methyl methacrylate) membranes as both plasticizers and as ion responsive media.[20] The electrodes exhibited "good and extremely" stable responses to cations and anions, including surfactant ions such as cetylpyridinium cations and dodecylsulfate anions.

In 2007, Safavi and co-workers reported the use of the imidazolium and pyridinium ionic liquids [C_4mim][PF_6] and [C_8py][PF_6], respectively, as replacements for the paraffin that is employed as a binder for the construction of carbon paste potentiometric sensors.[21] Carbon paste electrodes are robust and inexpensive. They are typically prepared by dispersing graphite powder in the paraffin. However, they do not contain an ionophore. Their potentiometric response depends on the partition of ions between the electrode and the aqueous phase containing the ions. Compared with traditional carbon paste electrodes, the carbon ionic liquid electrodes developed by Safavi and co-workers exhibited increased performance in terms of selectivity and potentiometric response times to Ag^+, Hg^{2+}, Fe^{3+} and other metal cations.

13.7 BIOSENSORS

Biosensors are devices that incorporate a biological component, often an enzyme, to detect an analyte. Over the past few years, there have been numerous reports of electrochemical biosensors based on electrodes modified with an ionic liquid to immobilize an enzyme.[22]

The electrodes are typically made of glassy carbon and coated with a composite material consisting of an ionic liquid and another material, such as carbon nanotubes, chitosan or clay. The composite material acts as a matrix for the enzyme. The biocatalytic activity of the enzyme towards a specific analyte is commonly determined by amperometry.

Reports of ionic liquid modified enzyme electrodes, notably by scientists in China, began to appear in 2005. Li and colleagues described an amperometric biosensor for detecting hydrogen peroxide that used a silica-based porous sol–gel hybrid material containing $[C_4mim][BF_4]$ to immobilize the enzyme horseradish peroxidise.[23] The group reported that the ionic liquid based sol–gel matrix exhibited much higher activity than that in the sol–gel matrix without the ionic liquid.

Several reports have focused on the use of composite materials containing an ionic liquid and carbon nanotubes. Dong and co-workers, for example, used a composite of multiwalled carbon nanotubes and $[C_4mim][PF_6]$ as a matrix to immobilize the enzyme glucose oxidase on a glassy carbon electrode.[24] The modified electrode showed good stability and exhibited good electrocatalytic activity for detecting glucose.

Ionic liquids might also prove useful as solvents for biosensor applications. Baker and colleagues have shown that the water-miscible ionic liquid $[C_4mim][BF_4]$ and other $[C_4mim]^+$ ionic liquids have potential as biosensor solvents for poorly water-soluble analytes such as fungicides, pesticides, environmental pollutants, fat-soluble metabolites and illicit drugs.[25]

The researchers carried out immunoanalyses that target a green fluorescent hapten dye known as BODIPY FL. The dye, which is a spectral analogue of fluorescein, was dissolved in aqueous solutions of the $[C_4mim]^+$ ionic liquids. The group used polyclonal rabbit anti-BODIPY FL antibodies as the biorecognition elements. When the antibodies bind to the dye molecules, the green fluorescence is strongly quenched. The team demonstrated that the antibodies bind with high affinity to the dye when they are dissolved in aqueous solutions of the ionic liquids and also when they are immobilized on solid supports and immersed in pure $[C_4mim]^+$ ionic liquids.

13.8 SPECTROSCOPY

Their capacity to dissolve a wide range of inorganic and organic materials makes ionic liquids suitable as solvent media for many conventional spectroscopic techniques such as ultraviolet/visible spectroscopy. For example, ionic liquids as solvents for UV/Vis spectroscopy have been used for various applications such as the study of metal complexes.[26]

Many of the studies over recent years have focused on molecular fluorescence spectrometry. In one of the earliest investigations in this field, a group led by Pandey investigated the fluorescence quenching of alternant and non-alternant polycyclic aromatic hydrocarbons (PAHs) by nitromethane in $[C_4mim][PF_6]$.[27] PAHs are carcinogens and mutagens that are released into the environment by the incomplete combustion of fossil fuels and biomass. Alternant PAHs are fully conjugated systems whereas the aromaticity of nonalternant PAHs is disrupted by five-membered rings. The alternant PAHs are efficiently quenched by nitromethane within $[C_4mim][PF_6]$ while non-alternant PAHs are not quenched at all. The quenching of the alternant PAHs in the ionic liquid was shown to be more efficient than that in a glycerol/water mixture with similar viscosity to the ionic liquid. The results suggest that ionic liquids may find use as solvents for PAH analysis and other areas of environmental analysis.

Magnetic resonance techniques such as ^1H nuclear magnetic resonance (NMR) spectroscopy, ^{13}C NMR, ^{17}O NMR, ^{31}P NMR, ^{35}Cl NMR, ^{27}Al NMR, ^{127}I NMR, nuclear Overhauser enhancement spectroscopy (NOESY), ^1H rotating frame Overhauser effect spectroscopy (ROESY) and electron paramagnetic resonance (EPR) spectroscopy have been used to study ionic liquids and chemistry in ionic liquid solvents.

Compared with molecular organic solvents, ionic liquids impose several limitations on conventional forms of NMR. ^1H NMR, for example, requires deuterated solvents whereas ionic liquids are not normally available in their deuterated forms. The viscosity of many ionic liquids is also a drawback. Finally, as ionic liquids are salts, they give

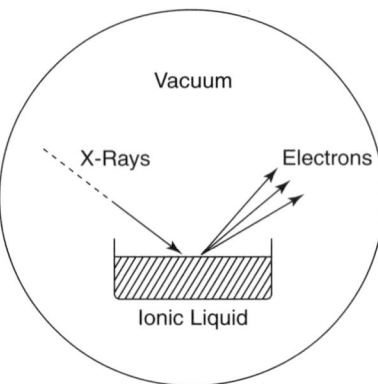

Figure 13.1 Schematic diagram of X-ray photoelectron spectroscopy of an ionic liquid.

rise to broad spectral lines. In many cases, sophisticated NMR techniques are required for the study of ionic liquids and their use as solvents. For further information on the use of NMR spectroscopy and EPR spectroscopy to study ionic liquids, the reader is referred to two extensive reviews of the topic.[28,29]

The lack of vapour pressure of ionic liquids means that they do not evaporate under vacuum and can therefore be used in ultra-high vacuum spectroscopic techniques. One such technique is X-ray photoelectron spectroscopy, which provides information about the electronic states of different kinds of atoms in a compound. The technique has sufficient sensitivity to distinguish between atoms of the same element located in chemically distinct environments in a compound. Licence and co-workers used the technique to study a commercial ionic liquid, [C$_2$mim][C$_2$H$_5$SO$_4$], and to monitor the reduction of Pd^{2+} ions in a solution of a palladium catalyst in the liquid (Figure 13.1).[30]

13.9 MASS SPECTROMETRY

The combination of ionic liquids and mass spectrometric techniques such as electrospray ionization mass spectrometry (ESI-MS) and matrix-assisted laser desorption/ionization mass spectrometry (MALDI-MS) has been used in several applications.

The electrospray process transfers analyte ions from a solution into the gas phase. The process can transfer large thermally fragile molecules that are highly polar – notably biomolecules – into the gas phase with little or no fragmentation. A volatile polar solvent that allows the formation of a stable spray, such as water or acetonitrile, is normally used. The process enables picomolar concentrations (10^{-12} mol dm^{-3}) of ions to be analysed by mass spectrometry.

The use of ESI-MS to analyse low concentrations of air- and water-sensitive ionic catalysts immobilized in ionic liquids after a reaction is potentially attractive. However, as ionic liquids have negligible vapour pressures, they do not evaporate. Furthermore, the high concentration of ionic liquid ions can swamp the analyte ions and interfere with the analysis. These problems can be overcome by diluting the ionic liquid with an organic solvent.

Dyson and colleagues showed that an ESI-MS technique known as electrospray ionization ion trap mass spectrometry can detect picomolar concentrations of ruthenium- and rhodium-containing cationic catalysts and the high molecular weight anionic catalyst $[HOs_3W(CO)_{14}]^-$ in $[C_4mim][PF_6]$ diluted with methanol.[31] The team also prepared a new ionic liquid with a catalyst anion, $[C_4mim][Rh(CO)_2I_2]$, and used the technique to characterize the ionic liquid, diluted to a catalytic amount in $[C_4mim][PF_6]$.

MALDI-MS is a mass spectrometric ionization technique widely used for the analysis of high molecular weight compounds such as proteins, oligonucleotides, sugars and synthetic polymers. The technique relies on a laser for ionization and the use of a vacuum-stable matrix, typically a non-volatile acidic aromatic compound. The analyte is mixed with a solution of the matrix and the solution spotted onto a plate. The solution is allowed to evaporate under vacuum, leaving the matrix containing the analyte. The matrix embeds and protects the analyte and also facilitates its ionization by provision of protons when a laser is directed at the spot. The matrix becomes desorbed along with intact molecular ions of the analyte during the process.

Ionic liquids have certain advantages as MALDI matrices over widely used solid and liquid matrices. For example, they allow the formation of homogeneous matrix–analyte solutions, are more stable in vacuums and give more reproducible results.

The first report of the use of ionic liquids as matrices for MALDI-MS appeared in 2001.[32] Armstrong and co-workers tested a wide range of ionic liquids for the analysis of peptides, proteins and poly(ethylene glycol). They showed that most conventional ionic liquids, such as $[C_4mim][PF_6]$ and $[C_4mim][BF_4]$, are ineffective as MALDI matrices because they do not make protons available for ionization of the analytes. However, ionic liquids with cations containing acidic protons, prepared from traditional acidic MALDI matrices, outperformed the parent solid matrices. For example, the ionic liquid matrix **13.2**, prepared from α-cyano-4-hydroxycinnamic acid (CHCA) and 1-methylimidazole, gave greater spectral peak intensities than the solid CHCA matrix when used for the analysis of bradykinin, a peptide consisting of a chain of nine amino acids.

13.2

REFERENCES

1. F. Pacholec, H. T. Butler and C. F. Poole, *Anal. Chem.*, 1982, **54**, 1938.
2. D. W. Armstrong, L. He and Y.-S. Liu, *Anal. Chem.*, 1999, **71**, 3873.
3. A. Berthod, L. He and D. W. Armstrong, *Chromatographia*, 2001, **53**, 63.
4. Y. Sun and A. M. Stalcup, *J. Chromatogr. A*, 2006, **1126**, 276.
5. W. Zhang, L. He, Y. Gu, X. Liu and S. Jiang, *Anal. Lett.*, 2003, **36**, 827.
6. A. Berthod, M.-J. Ruiz-Angel and S. Carda-Broch, in *Ionic Liquids in Chemical Analysis*, ed. M. Koel, CRC Press, Boca Raton, FL, 2009, p. 211.
7. E. G. Yanes, S. R. Gratz, M. J. Baldwin, S. E. Robison and A. M. Stalcup, *Anal. Chem.*, 2001, **73**, 3838.
8. R. Kuldvee, M. Vaher, M. Koel and M. Kaljurand, *Electrophoresis*, 2003, **24**, 1627.
9. M. Vaher and M. Koel, *J. Chromatogr. A*, 2005, **1068**, 83.
10. S. M. Mwongela, A. Numan, N. L. Gill, R. A. Agbaria and I. M. Warner, *Anal. Chem.*, 2003, **75**, 6089.
11. W. Qin and S. F. Y. Li, *J. Chromatogr. A*, 2004, **1048**, 253.
12. C. Liang, C.-Y. Yuan, R. J. Warmack, C. E. Barnes and S. Dai, *Anal. Chem.*, 2002, **74**, 2172.
13. R. E. Baltus, B. H. Culbertson, S. Dai, H. Luo and D. W. DePaoli, *J. Phys. Chem. B*, 2004, **108**, 721.
14. R. Wang, T. Okajima, F. Kitamura and T. Ohsaka, *Electroanalysis*, 2004, **16**, 66.
15. R. Wang, S. Hoyano and T. Ohsaka, *Chem. Lett.*, 2004, **33**, 6.

16. D. S. Silvester, K. R. Ward, L. Aldous, C. Hardacre and R. G. Compton, *J. Electroanal. Chem.*, 2008, **618**, 53.
17. L. E. Barrosse-Antle, D. S. Silvester, L. Aldous, C. Hardacre and R. G. Compton, *J. Phys. Chem. C*, 2008, **112**, 3398.
18. A. M. O'Mahony, D. S. Silvester, L. Aldous, C. Hardacre and R. G. Compton, *J. Phys. Chem. C*, 2008, **112**, 7725.
19. C. Coll, R. H. Labrador, R. M. Mañez, J. Soto, F. Sancenón, M.-J. Seguí and E. Sanchez, *Chem. Commun.*, 2005, 3033.
20. N. V. Shvedene, D. V. Chernyshov, M. G. Khrenova, A. A. Formanovsky, V. E. Baulin and I. V. Pletnev, *Electroanalysis*, 2006, **18**, 1416.
21. A. Safavi, N. Maleki, F. Honarasa, F. Tajabadi and F. Sedaghatpour, *Electroanalysis*, 2007, **19**, 582.
22. S. N. Baker, T. A. McCarty, F. V. Bright, W. T. Heller and G. A. Baker, in *Ionic Liquids in Chemical Analysis*, ed. M. Koel, CRC Press, Boca Raton, FL, 2009, p. 99.
23. Y. Liu, L. Shi, M. Wang, Z. Li, H. Liu and J. Li, *Green Chem.*, 2005, **7**, 655.
24. Y. Liu, L. Liu and S. Dong, *Electroanalysis*, 2007, **19**, 55.
25. S. N. Baker, E. B. Brauns, T. M. McCleskey, A. K. Burrell and G. A. Baker, *Chem. Commun.*, 2006, 2851.
26. M. Koel, in *Ionic Liquids in Chemical Analysis*, ed. M. Koel, CRC Press, Boca Raton, FL, 2009, p. 295.
27. K. A. Fletcher, S. Pandey, I. K. Storey, A. E. Hendricks and S. Pandey, *Anal. Chim. Acta*, 2002, **453**, 89.
28. D. Bankmann and R. Giernoth, *Progr. Nucl. Magnet. Res. Spectrosc.*, 2007, **51**, 63.
29. R. Giernoth, in *Ionic Liquids in Chemical Analysis*, ed. M. Koel, CRC Press, Boca Raton, FL, 2009, p. 355.
30. E. F. Smith, I. J. V. Garcia, D. Briggs and P. Licence, *Chem. Commun.*, 2005, 5633.
31. P. J. Dyson, J. S. McIndoe and D. Zhao, *Chem. Commun.*, 2003, 508.
32. D. W. Armstrong, L.-K. Zhang, L. He and M. L. Gross, *Anal. Chem.*, 2001, **73**, 3679.

CHAPTER 14
Applications

14.1 INTRODUCTION

Room-temperature ionic liquids have several properties that have attracted industrial interest throughout the world. These properties include thermal and electrochemical stability, conductivity, low vapour pressure, non-flammability, tunability, a broad range of miscibilities and wide liquid temperature ranges. The possibility of using room-temperature ionic liquids to replace volatile organic solvents is one of the major attractions.

Earlier chapters in this book outlined various potential applications. The use of ionic liquids for separations and extractions, for example, is described in the chapter on designer solvents (Chapter 5). The possibility of exploiting ionic liquids as electrolytes for batteries and other applications dates back to the early 1960s (Chapter 7). In organic synthesis ionic liquids are particularly useful in biphasic catalysis for immobilizing, activating and recycling catalysts (Chapter 8). They can also have a pronounced influence on the yield of and selectivity for a desired product. Many biotransformations can also be carried out in ionic liquids (Chapter 12). Finally, in analysis, ionic liquids have proven useful as stationary phases for gas and liquid chromatography, and for capillary electrophoresis and gas sensors (Chapter 13).

The extent to which industrial companies develop processes that use ionic liquids, or develop ionic liquids as materials for various applications, will depend not only on their chemical, physical and biological properties but also on several other factors, including their cost,

An Introduction to Ionic Liquids
By Michael Freemantle
© Michael Freemantle 2010
Published by the Royal Society of Chemistry, www.rsc.org

availability, green credentials such as recyclability, and the possibility of retro-fitting ionic liquid-based processes into existing plants that use conventional solvents.

The following selection of past, present and potential applications of ionic liquids, both as solvents and as materials, illustrates the scope and commercial potential of these liquids.

14.2 SYNTHESIS OF 2,5-DIHYDROFURAN

In 1996, the Eastman Chemical Company launched a process that employed an ionic liquid as a co-catalyst for the isomerization of 3,4-epoxy-1-butene to 2,5-dihydrofuran (Scheme 14.1).[1,2] This compound is used as an intermediate for the manufacture of many commodity, speciality and fine chemicals such as tetrahydrofuran and other types of furans.

The isomerization of 3,4-epoxy-1-butene to 2,5-dihydrofuran has been known since 1976. The rearrangement is carried out homogeneously in the liquid phase using a metal halide Lewis acid and an iodide salt as co-catalysts. The original process used a high-boiling solvent and resulted in side reactions such as the polymerization of 3,4-epoxy-1-butene to non-volatile polyether oligomers. These by-products diluted the catalysts and filled up the reaction vessel. Recovery of the catalysts and solvent proved difficult.

Eastman Chemical therefore searched for an inexpensive miscible co-catalyst system that exhibited good stability, activity and selectivity for 2,5-dihyrofuran, as well as low toxicity and good recoverability. In addition, it was necessary that the catalysts were non-volatile so that they did not co-distil with the product.

The process developed by the company employs a trialkyltin iodide, which is a Lewis acid catalyst, and a Lewis basic tetraalkylphosphonium iodide ionic liquid, [P$_{8\ 8\ 8\ 18}$]I, as a co-catalyst. The phosphonium ionic liquid was preferred to a tetraalkylammonium iodide ionic liquid because of its greater thermal stability. The catalysts were recovered by liquid–liquid extraction.

The continuous liquid-phase process was operated by Texas Eastman Division in a plant at Longview, Texas, with an annual capacity of 1400

Scheme 14.1 Isomerization of 3,4-epoxy-1-butene to 2,5-dihydrofuran.

metric tons from December 1996 to December 2004. The plant was then shut down because the market for furan products had declined.

14.3 DIFASOL PROCESS

In March 1998, the French Petroleum Institute (IFP) at Rueil-Malmaison, near Paris, launched a commercial process, known as the Difasol process, that employs ionic liquids for the production of iso-octenes from butenes.[3] Isooctenes are hydroformylated in industry to produce alcohols such as isononanols. Esterification of the alcohols yields dialkyl phthalates, which are used as plasticizers for poly(vinyl chloride) (PVC).

The Difasol process dimerizes n-butene in a continuous two-phase liquid–liquid operation that is carried out between −15 and 0 °C. It employs a catalyst dissolved in an acidic chloroaluminate ionic liquid consisting of a mixture of $AlCl_3$ and either dichloroethylaluminium, $C_2H_5AlCl_2$, or chlorodiethylaluminium, $(C_2H_5)_2AlCl$.[4] The olefin is continuously fed into the reactor. The isooctene products are poorly miscible with the ionic liquid and separate out in a settling tank.

The catalyst used for the process is a nickel salt that is activated by alkylaluminium chloride compounds. It is the same catalyst as that used for IFP's Dimersol processes for dimerizing alkenes. The Dimersol processes are homogeneous processes that do not involve ionic liquids. The processes are used world wide and produce millions of tonnes of the dimer products. They include the Dimersol-G process for propene dimerization and the Dimersol-X process for butene dimerization.

The Difasol ionic liquid process for manufacturing isooctenes consumes less catalyst than the non-ionic liquid Dimersol-X process. The volume of the Difasol reactor is also smaller than that used for the Dimersol-X process. The Difasol ionic liquid–catalyst system not only leads to better butene conversions compared with the Dimersol-X process but also higher selectivity for the dimers rather than for longer-chained oligomers. The biphasic system has the additional advantage of enabling the catalyst to be recycled efficiently.

The possibility of retro-fitting the Difasol process into existing Dimersol plants is an additional advantage. The biphasic Difasol process can also be used in combination with the homogeneous Dimersol-X process. Unconverted butenes from the Dimersol-X reactor are condensed from the vapour phase and sent to the Difasol reactor. The combination improves the yield of isooctenes by about 25% and also consumes less catalyst. Furthermore, the butenes fed into the Difasol reactor are completely pure. As a consequence, accumulation of impurities in the ionic liquid phase is avoided.

14.4 BASIL PROCESS

In early 2002, BASF commenced routine operation of its "Biphasic Acid Scavenging utilising Ionic Liquids" (BASIL) process at its site in Ludwigshafen, Germany.[5] The process produces alkoxyphenylphosphines, which are used as precursors for the synthesis of photoinitiators used in the manufacture of printing inks.

The alkoxyphenylphosphines are formed by the reaction of chlorophenylphosphines with alcohols. Acid is also generated during the reaction. In an earlier non-ionic liquid process, triethylamine was used as a base to scavenge the acid. The reaction resulted in the precipitation of the insoluble waste by-product triethylammonium chloride. The by-product had to be removed by filtration.

The BASIL process employs 1-methylimidazole to scavenge the acid. The protic ionic liquid 1-methylimidazolium chloride, [Hmim]Cl, which has a melting point of 75 °C, is formed by the reaction (Scheme 14.2). The ionic liquid separates out as a clear liquid phase from the pure product and is recycled back to 1-methylimidazole by deprotonation with a base. The generation of the ionic liquid is therefore switched on by protonation of the 1-methylimidazole and switched off by deprotonation. Once switched off, the 1-methylimidazole can be removed by distillation like other organic solvents.

Although 1-methylimidazole is a weaker base than triethylamine, it is a more efficient acid scavenger. The original triethylamine process had to be run below 30 °C to achieve reasonable yields of the alkoxyphenylphosphines. The BASIL process is run at 90 °C with increased

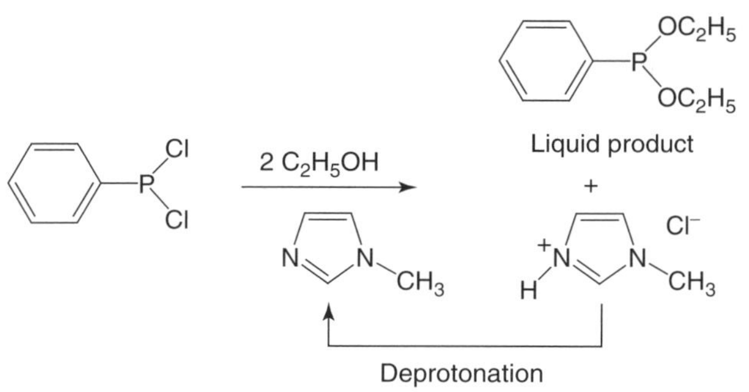

Scheme 14.2 BASIL process.

yields and the liquid–liquid biphasic system facilitates continuous processing. In addition, 1-methylimidazole is a nucleophilic catalyst that accelerates the reaction.

14.5   STORAGE OF HAZARDOUS GASES

In 2005, Tempel and co-workers at Air Products, Allentown, USA, revealed that the company had developed systems based on ionic liquids for the storage and delivery of two gases: phosphine (PH_3) and boron trifluoride (BF_3).[6] The ionic-liquid gas storage technology, commercially named GASGUARD Sub-Atmospheric Systems, provides a safe and effective way of storing and delivering the two gases in high purity to the electronics markets.

Many gases used in industry are hazardous because of their toxicity, flammability and reactivity. Hazardous gases are commonly stored at high pressure in cylinders either as pure gases or as part of a mixture with an inert gas. Compared with storage of hazardous solids and liquids that have negligible vapour pressures, high-pressure storage of gases poses a significant risk. An unintended or uncontrollable release can result in significant injury or death.

PH_3, BF_3 and arsine (AsH_3) are examples of hazardous gases. They are highly toxic. The three gases are used in the electronics industry to dope silicon wafers with phosphorus, boron and arsenic ions, respectively. Doped silicon is an important component of the field-effect transistors that are used extensively to fabricate the microprocessors used in computers, cellular phones and other devices.

The ion implanter, the tool that implants ions into silicon, operates at high voltages. The gas source, which must be at the same voltage, has to be installed within the ion implanter and be located close to its operators. The preferred commercial ion implantation technology nowadays stores the gases by physical adsorption onto activated carbon or a zeolite inside a standard gas cylinder. The gases are stored at less than ambient pressure and delivered to the ion implantation process under vacuum on demand. This method sidesteps the safety risk of storing toxic gases at high pressures and also allows a large quantity of gas to be stored in a small volume at less than ambient pressure.

The Air Products GASGUARD technology relies on the ability of ionic liquids to store gases reversibly through chemical complexation rather than physical adsorption. Tempel and colleagues showed that PH_3, which is a relatively weak Lewis base, complexes reversibly with Lewis acidic ionic liquids. They found that the best results were obtained with the ionic liquid [C$_4$mim][Cu$_2$Cl$_3$]. The PH$_3$ complexes with the ionic liquid in the molar ratio 2:1.[7]

The researchers subsequently extended their investigations to BF$_3$, a strongly Lewis acidic gas. They first used molecular modelling to predict that the gas would reversibly complex with basic dialkylimidazolium tetrafluoroborate salts in equimolar ratios. Subsequent laboratory experiments confirmed the prediction. Experimental evidence and theoretical results suggested that the BF$_3$ molecule complexes with the anion of ionic liquids such as [C$_4$mim][BF$_4$] to form the adduct [C$_4$mim][BF$_3$-F-BF$_3$].

The gas storage capacities of [C$_4$mim][Cu$_2$Cl$_3$] and [C$_4$mim][BF$_4$] are "exceptionally high", according to Tempel and co-authors. The use of these ionic liquids is also attractive as the liquids can be readily agitated mechanically and pumped. In addition, the ionic liquids are better than porous solids at transferring heat. They can also be recycled numerous times without losing their capacity to store the gases.

The new system enables PH$_3$ or BF$_3$ to be stored in cylinders containing the ionic liquids and transported at pressures near atmospheric pressure. This is far less than the pressures typically required for storing and shipping these poisonous gases. Pressurized cylinders are therefore not required.

To deliver the PH$_3$ or BF$_3$, the cylinder is simply connected to a vacuum source. Reducing the pressure reverses the chemical complexation process and frees the reactive gas.

14.6 LUBRICATION

In 2001, Liu and colleagues recognized that ionic liquids have a range of properties that might make them useful as lubricants.[8] The properties include negligible vapour pressure, lack of flammability, high thermal stability and fluidity at low temperatures. Low flammability, for example, is a desirable property for the lubricants used to lubricate machines in confined spaces such as mines.

Liu's group demonstrated that dialkylimidazolium tetrafluoroborates are potentially versatile lubricants for various ceramic and metal surfaces. The team determined the tribological behaviour of 1-hexyl-3-methylimidazolium tetrafluoroborate, [C$_6$mim][BF$_4$], and its 3-ethyl analogue using a standard friction- and wear-testing technique for frictional pairs such as steel/steel, steel/copper, steel/SiO$_2$ and silicon nitride/SiO$_2$. The two ionic liquids proved to have "excellent" friction reduction and antiwear properties for lubricating steel or Si$_3$N$_4$ that is in contact with other metals and ceramics.

The two ionic liquids are non-hygroscopic, air- and water-stable at room temperature, and slightly soluble in water. The addition of a small amount of water to the liquids was found to improve the anti-wear ability of the frictional pairs steel/steel, steel/aluminium and steel/ceramic. The addition of water, however, had little effect on reducing the friction.

Because of their high thermal stabilities, the possibility of using ionic liquids as high-temperature lubricants for use in aircraft engines is also being explored. Conventional lubricants used in aircraft engines only function at temperatures up to 150 °C. Engines in advanced military aircraft, on the other hand, require lubricants that function reliably for over 4000 hours from –40 to over 300 °C.

Shreeve and co-workers have prepared several ionic liquids that exhibit high thermal stability and good lubricity at 300 °C.[9] The ionic liquids are dicationic. They typically consist of two imidazolium cations linked by poly(ethylene glycol) and functionalized by alkyl or polyfluoroalkyl groups (**14.1**). The decomposition temperatures of the dicationic ionic liquids with alkyl substituents are higher than those with polyfluoroalkyl groups. On the other hand, the ionic liquids with polyfluoroalkyl groups exhibit better antiwear properties. In general, the dicationic ionic liquids exhibit "excellent" tribological characteristics and are potentially suitable as high temperature lubricants, according to the researchers.

14.1

Although ionic liquids generally have low coefficients of friction and other desirable lubricating properties, they require additives for optimal lubrication just as they are required for the conventional oils used as lubricants.[10]

14.7 COMPRESSORS

Scientists at Linde Gas in Vienna have developed an "ionic compressor" that uses a custom-made ionic liquid to replace the metal pistons used in conventional compressors.[11] The ionic liquid, which is barely compressible, acts as a liquid piston in a cylinder. The up-and-down motion of the liquid compresses the gas. Because the ionic liquid is immiscible with the gas, the compressor does not require seals, bearings or lubrication. As a result, the number of moving parts is reduced from around 500 in a conventional compressor to eight. In addition, the ionic compressor makes little noise compared with that of conventional compressors.

The Linde technology has been used to fuel a fleet of natural gas powered cars since July 2005. It is also being developed for use in hydrogen fuelling stations. Charging airbags with compressed gases is another potential application.

14.8   ENERGETIC MATERIALS AND PROPELLANTS

Ionic liquids with nitrogen-containing cations and oxidizing anions are potentially attractive high energy density materials for industrial and military applications such as explosives and propellants for driving turbines or propelling rockets. High energy density materials are materials that pack a lot of energy into a small volume.

Many propellants combine a fuel and an oxidizer. Rockets, for example, are propelled by the gases that rapidly evolve when the rocket fuel is oxidized. Ammonium nitrate is an example of a propellant. It combines ammonium as the fuel and nitrate as the oxidizer. When detonated, the compound decomposes explosively to form gaseous nitrogen and oxygen, and water vapour. The salt, however, is a solid. Consequently, once the compound has been detonated, it is difficult to turn off the propulsion. All the propellant therefore has to be consumed in a single firing.

Liquid propellants, in contrast, can be turned on and off like a tap. They can be either monopropellants or bipropellants. Monopropellants combine the fuel and oxidant in a single compound. Bipropellants combine a fuel, such as kerosene, and an oxidizer such as liquid oxygen. Bipropellants require heavier and more complex equipment than their monopropellant counterparts. For example, two tanks are needed to store the fuel and oxidizer separately. In addition two pumps and flow controllers are required.

Although the performance of liquid monopropellants is normally much lower than that of solid monopropellants and liquid bipropellants, they are simpler to use and, by firing them for short periods, they can readily stabilize and adjust the movement of rockets and spaceships.

Hydrazine, N_2H_4, was used as a liquid monopropellant for rocket-powered engines used in fighter planes in World War II. It continues to be used for spacecraft. In hydrazine engines, the monopropellant passes over a transition metal catalyst and decomposes exothermically into gaseous nitrogen, hydrogen and ammonia. However, the compound suffers from two distinct disadvantages. It has a high vapour pressure and is carcinogenic.

Propellants, whether solid or liquid, should possess high energy density and also be rich in nitrogen so that they produce nitrogen gas when oxidized. Nitrogen gas is benign. It does not pollute the environment and is not a risk to health. The ideal liquid propellant should also have a low melting point, negligible vapour pressure, low viscosity, minimal shock and friction sensitivities, and high thermal and hydrolytic stabilities. In addition, it should also be non-toxic and, to avoid pollution, have a low carbon content.

Over recent years, several research groups have been searching for high energy density ionic liquids that combine some or all of these features and might therefore be used as propellants. Shreeve and co-workers, for example, have synthesized and characterized various ionic liquids consisting of nitrogen-rich fuel cations and oxidizing anions.[12] They include energetic salts of alkylimidazolium (**14.2**), alkyltriazolium (**14.3**) or alkylpyridinium (**14.4**) cations substituted with a pentafluorosulfanyl SF_5 group, and highly oxidizing anions such as nitrate, $[NO_3]^-$, or dinitramide, $[N(NO_2)_2]^-$.[13] The salts are substituted with SF_5 to improve the densities and thermal stabilities of the materials. Most of the salts have melting points below $100\,^\circ C$, good thermal stabilities and moderately high densities.

14.2   **14.3**

14.4

$R = CH_2CH_2CH_2SF_5$

One of the key considerations for the development of energetic ionic liquids is oxygen balance. The oxidizing anions should contain sufficient oxygen for complete oxidation of the relatively large fuel cations.

Christe and co-workers have synthesized an oxygen-balanced energetic ionic liquid that might potentially be used as a liquid monopropellant.[14] The ionic liquid, which has a glass transition temperature of $-46\,^\circ C$, consists of an energetic 1-ethyl-4,5-dimethyltetrazolium cation and a complex oxidizing anion: tetranitratoaluminate (**14.5**). The anion contains 12 oxygen atoms, 10.5 of which are available to oxidize the fuel cation. The high oxygen content of the anion allows complete combustion of the large organic cation. The performance of the compound as a liquid monopropellant "significantly exceeds those of state-of-the art materials, such as hydrazine", according to the authors.

14.5

Ionic liquids with energetic anions based on nitrogen heterocyclic compounds are also known. Katritsky and co-workers, for example, synthesized a series of ionic liquids with energetic azolate anions.[15] They prepared 28 salts by combining four cations – $[C_4mim]^+$ and three tetraalkylammonium cations – with seven heterocyclic anions, including 2,4-dinitroimidazolate (**14.6**), 3,5-dinitro-1,2,4-triazolate (**14.7**) and tetrazolate (**14.8**).

14.6   **14.7**   **14.8**

14.9   OPTICAL IMMERSION FLUIDS

Optical immersion fluids are used to examine inclusions in minerals, such as quartz, and gems, such as diamond. Inclusions are foreign materials – small amounts of gases, liquids or other minerals – that became entrapped in the host minerals or gems when they crystallized in the Earth's interior. These enclosed "impurities" or "flaws" provide mineralogists with valuable information about the genesis of minerals and gems and their chemical and physical environment at the time of formation.

The inclusions can be examined without destroying the host gem or mineral samples by use of an optical microscope. The samples are immersed in an optical immersion fluid with a refractive index that matches that of the sample. The fluid eliminates interference of light reflected from the surface of the sample.

Refractive indices frequently uniquely identify a mineral or gem. However, they tend to be higher than readily available fluids such as water and ethanol. These liquids have refractive indices in the range 1.30 to 1.40 whereas diamond, for example, has a unique refractive index of 2.42. Many of the high refractive index immersion fluids used nowadays are solid at room temperature and also poisonous and unstable. They include arsenic(III) iodide, AsI_3, and tin(IV) iodide, SnI_4, which have refractive indices of 2.2 and 2.1, respectively.

Ionic liquids with high refractive indices are promising alternatives as optical immersion fluids for optical mineralogy studies. Because of their lack of vapour pressure, they are more environmentally benign than the noxious optical immersion fluids that are currently available.

Deetlefs and co-workers have synthesized 20 ionic liquids with refractive indices in the range 1.40–2.10.[16] The ionic liquids have 1-alkyl-3-methylimidazolium cations. The refractive indices of ionic liquids with these cations and the same anion decrease with increasing length of the alkyl chain in the cation.

By mixing the 20 ionic liquids in simple 1 : 1 volume ratios, 400 binary mixtures can be prepared. Even more mixtures can be prepared by mixing the pure components in different ratios. The pure liquids and binary mixtures can then be used as optical immersion fluids with refractive indices ranging from 1.40 to 2.10 in intervals of 0.01.

14.10 LUNAR LIQUID MIRROR TELESCOPE

The possibility of using an ionic liquid to form the base for the reflective metal surface of a liquid mirror telescope stationed on the Moon is being explored by an international team of researchers.

The infrared Lunar Liquid Mirror Telescope concept, proposed by the National Aeronautics and Space Administration (NASA) in the USA, has a large mirror some 20–100 metres in diameter. The proposed mirror consists of a spinning liquid that rotates to form a bowl shape, known as a parabola. The liquid has a thin metal film on its surface. When the liquid spins in a perfect parabola, the film reflects infrared light from distant stars and galaxies that cannot be picked up by telescopes on Earth because of atmospheric interference and light pollution.

Telescopes with parabolic liquid mirrors are less expensive and easier to make and maintain than conventional telescopes with glass mirrors. Liquid mirror telescopes employed in observatories on Earth traditionally use mercury as the reflective liquid. However, mercury cannot be used for a lunar liquid mirror telescope as the high-vacuum conditions on the Moon would cause the mercury to boil.

The lunar liquid telescope requires a liquid with low vapour pressure that also does not freeze at the sub-zero temperatures found on the Moon. The liquid must also have high viscosity and remain stable indefinitely.

As ionic liquids have negligible vapour pressures, they generally do not boil, even under vacuum. Many have very low melting points and are stable and highly viscous. Ionic liquids are also much lighter than mercury – a key consideration for transporting a telescope to the Moon.

The international team has shown, in a proof-of-concept study, that 1-ethyl-3-methylimidazolium ethylsulfate, [C_2mim][$C_2H_5OSO_3$], a commercially-available ionic liquid that possesses these properties, can be coated with a nanolayer of chromium followed by a reflective nanolayer of silver.[17] The reflectivity of a coating of silver deposited on chromium is better than that of a coating of silver alone.

[C_2mim][$C_2H_5OSO_3$] remains liquid at temperatures down to $-98\,°C$. The lunar liquid mirror telescope, however, will need a liquid with a melting point as low as $-173\,°C$. Fortunately, some one million or so simple ionic liquids are theoretically possible and it should, in principle, be possible to design an ultra-low melting point ionic liquid or a binary or ternary ionic liquid system with an ultra-low eutectic point and the other properties required for the telescope.

14.11 THERMOMETERS

In 2008, a team of scientists from Europe and the USA described a prototype liquid-in-glass thermometer that employs an ionic liquid.[18]

Most commercial liquid-in-glass thermometers use mercury or ethanol as filling fluids. Mercury is volatile and toxic, even at low concentrations, and is therefore potentially harmful if allowed to escape into the environment. It freezes at $-39\,°C$ which is higher than some of the low temperatures that occur in the polar regions of the Earth. Ethanol is also volatile but boils at $78.8\,°C$, which is below the boiling point of water. The range of applications for both types of thermometer is therefore limited.

The team selected two commercially-available ionic liquids, tris(2-hydroxyethyl)methylammonium methyl sulfate, [TEMA][CH$_3$SO$_4$] (**14.9**) and trihexyl(tetradecyl)phosphonium bis{(trifluoromethyl)sulfonyl}amide, [P$_{6\ 6\ 6\ 14}$][NTf$_2$], (**14.10**). Both salts are liquid over a wide range of temperatures and respond faster to temperature changes than mercury.

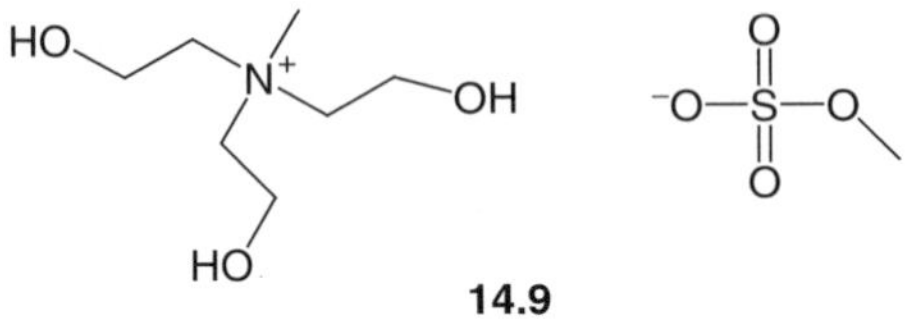

14.9

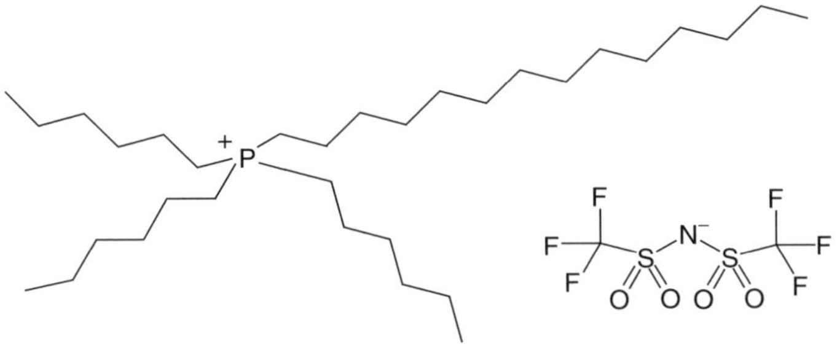

14.10

The thermometers could potentially be used in industry and in research and development laboratories to measure temperatures over conventional temperature ranges. They might also prove useful in environments with extremes of temperature, such as Antarctica and hydrothermal vents under the sea, where they would be able to measure temperatures lower than the freezing point of mercury and higher than the boiling points of ethanol and water.

For general use, the team selected [TEMA][CH$_3$SO$_4$] as it is reasonably inexpensive and a complete set of toxicological data is available. The ionic liquid remains stable from –81 to about 200 °C. For speciality uses in extremes of temperature, the team suggested the use of the phosphonium ionic liquid [P$_{6\ 6\ 6\ 14}$][NTf$_2$] as it has a glass temperature of –76 °C and is thermally stable up to 400 °C.

The prototype thermometers were constructed from thick-wall Pyrex tubes with narrow internal diameters and bulbs at one end that act as reservoirs for the ionic liquids. Both ionic liquids used in the thermometers are colourless. The team therefore added a red ionic liquid dye to the fluids to make them visible.

14.12 ANTIMICROBIAL AGENTS

The possibility of exploiting ionic liquids with antimicrobial activity as antiseptics, disinfectants and anti-fouling agents has triggered several investigations since the early 2000s.

In 2003, Pernak and colleagues described a study of the antimicrobial activity of a range of 1-alkoxymethyl-3-methylimidazolium ionic liquids with chloride, $[BF_4]^-$ and $[PF_6]^-$ anions.[19] They investigated the action of these ionic liquids against five strains of bacteria with spherical shapes, known as cocci, four strains of bacteria with rod shapes, and two strains of fungi. The strains included *Staphylococcus aureus*, the rod-shaped bacterium *Escherichia coli* (*E. coli*) and the fungus *Candida albicans*.

For each strain, the group determined the minimum inhibitory concentrations (MICs) and minimum biocidal (bactericidal or fungicidal) concentrations (MBCs) of the ionic liquids. MICs and MBCs are the minimum concentrations of antimicrobial agents that, respectively, inhibit the growth of or kill a microorganism. Low MIC and MBC values indicate high potency of an antimicrobial agent.

Pernak's results revealed a relationship between the structure of the ionic liquid's cation and antimicrobial activity. The most active ionic liquids were those with alkoxyl groups that contain 10, 11, 12 or 14 carbon atoms. Those with short substituents, *e.g.* propoxymethyl or butoxymethyl, were not active against bacteria and fungi.

Two years later, Docherty and Kulpa made the same observation for alkylimidazolium and alkylpyridinium ionic liquids.[20] The antimicrobial activities of the ionic liquids were greater for ionic liquids with cations having longer alkyl chain lengths. Similarly, Pernak and Feder-Kubis showed that ionic liquids with chiral alkylammonium cations with more than five carbon atoms in the alkyl group display a wide range of antimicrobial activities.[21]

Many pathogenic and environmental microorganisms grow in structured communities that are attached to the surfaces of solid substrates. The communities generate a polysaccharide matrix outside their cells, known as a glycocalyx. The polymeric coatings enable the communities to attach to surfaces and also protect them, *e.g.* against antibiotics and disinfectants. These encapsulated microbial communities are known as biofilms. The National Institutes of Health in the US has estimated that biofilms play a key role in up to 80% of all chronic human infections and that virtually all bacteria in aquatic ecosystems live in biofilm communities. In addition, microbial biofilms can also result in pipe blockages in industrial processes that use water.

In 2009, Gilmore and co-workers reported that 1-alkyl-3-methylimidazolium chloride ionic liquids are active against biofilms of clinically significant microbial pathogens, including MRSA and fungi.[22]

Methicillin-resistant *Staphylococcus aureus* (MRSA), also known as multiple-resistant *Staphylococcus aureus*, is a strain of the *Staphylococcus aureus* bacterium that is resistant to penicillins and related antibiotics. The strain is difficult to treat clinically and has caused numerous chronic infections and deaths, notably in hospitals.

Gilmore's team initially screened 1-alkyl-3-methylimidazolium chloride ionic liquids with alkyl chains lengths ranging from 4 to 18 carbon atoms for antimicrobial activity against MRSA. [C_4mim]Cl displayed no measurable activity while [C_{16}mim]Cl and [C_{18}mim]Cl were insufficiently soluble in the MRSA growth media.

The researchers selected 1-alkyl-3-methylimidazolium chloride ionic liquids with alkyl chain lengths ranging from 6 to 14 carbon atoms for further evaluation. They determined their MIC and MBC values against strains of MRSA and other bacteria, as well one fungal strain. The results showed that ionic liquids with alkyl chain lengths of more than ten carbon atoms exhibit potent, broad spectrum antimicrobial activity.

The group then grew biofilms of the pathogens and determined the antibiofilm activity of the ionic liquids in terms of their minimum biofilm eradication concentrations (MBECs). As with the MIC and MBC values, the MBEC values decreased with increasing alkyl chain length. The lowest values, and therefore the greatest antibiofilm activities, for all the microbial biofilms tested, including MRSA, were observed for [C_{14}mim]Cl.

14.13 ACTIVE PHARMACEUTICAL INGREDIENTS

One of the attractions of ionic liquids is the opportunity to tailor their physical, chemical and biological properties by building specific features into the chemical structures of their cations and independently into their anions. This dual nature of ionic liquids paves the way for the development of modular ionic liquids-based strategies that allow for the compartmentalized design of new pharmaceuticals with tunable biological properties, as well as tunable physical and chemical properties.

Many active pharmaceutical ingredients fail because they lack suitable chemical and physical properties such as solubility and stability. The development of active pharmaceutical ingredients based on ionic liquids should, in theory at least, allow not only control of the solubility, stability, bioavailability and bioactivity of ingredients but also control of the speed at which they are delivered to targets inside the body. In principle, it should be possible to design and prepare pharmaceutical ionic liquid cocktails containing various active pharmaceutical ingredients. The liquid cocktails could even be customized to the specific needs of a patient.

In 2007, an international team of 12 researchers described experiments that point to the potential power of such an ionic liquids-based strategy for the development of active pharmaceutical ingredients.[23] They prepared a room-temperature ionic liquid called lidocaine docusate (**14.11**) from two solid salts: lidocaine hydrochloride and sodium docusate. Lidocaine hydrochloride is a pain reliever that is used as a local surface anaesthetic in dentistry and for the treatment of neuralagia. Sodium docusate is a surfactant that is used mainly as an emollient in the treatment of constipation but also for the topical administration of drugs to the skin. Preliminary studies indicate that the ionic liquid lidocaine docusate combines and enhances the analgesic and emollient properties of the two solid salts.

14.11

The ionic liquids based strategy could also lead to more environmentally-friendly ways of manufacturing pharmaceuticals. Pharmaceuticals are currently manufactured in extensive synthetic procedures involving usage of large quantities of solvents and chemicals that end up as waste. Much of the chemical wastes are derived from attempts to improve the pharmaceutical properties such as solubility or bioavailability. The ionic liquids strategy, which is currently at its early stages of development, aims to eliminate much of this waste while at the same time delivering improved pharmaceutical performance.

14.14 NUCLEOSIDE-BASED ANTIVIRAL DRUGS

Nucleoside-based antiviral drugs are used to treat viral infections such as the human immunodeficiency virus (HIV), herpes simplex virus

(HSV) and the hepatitis B virus. The search for novel antiviral drugs based on natural nucleosides is often hampered by the poor solubility of nucleosides in conventional organic solvents. Nucleosides are more soluble in pyridine, dimethyl formamide (DMF) and dimethyl sulfoxide (DMSO), but these organic solvents are toxic.

Nucleoside chemists are also confronted with another problem. Nucleosides consist of two components, a base and a sugar. It is often necessary to protect one of these functionalities so that chemistry can be carried out on the other. Benzoylation in DMF or DMSO is commonly used for protection but selective benzoylation of the base amine or the sugar hydroxyl groups is difficult. Consequently, lengthy chromatographic procedures are needed to separate the desired benzoylated product.

In 2005, Prasad and co-workers reported that the ionic liquid 1-methoxyethyl-3-methylimidazolium methanesulfonate, $[CH_3OCH_2CH_2mim]$ $[CH_3SO_3]$, can be employed as a "green" alternative for the efficient and selective benzoylation of the hydroxyl groups of both ribonucleosides and deoxyribonucleosides under mild conditions.[24] The group selected the ionic liquid because the cation has a "sugarphilic" ether side chain.

Three of the same team subsequently showed that ribo- and deoxyribonucleosides are even more soluble in the related ionic liquid 1-methoxyethyl-3-methylimidazolium trifluoroacetate, $[CH_3OCH_2CH_2\text{-}mim][CF_3CO_2]$ (**14.12**). The ionic liquid is also an efficient reaction medium for the selective benzoylation of the sugar hydroxyl groups of the nucleosides.[25] They obtained high yields of O-benzoylated nucleosides under ambient conditions.

14.12

In 2008, Kumar and Malhotra described the synthesis of three nucleoside-based antiviral drugs in ionic liquids.[26] The drugs are Brivudine [also known as (*E*)-5-(2-bromovinyl)-2′-deoxyuridine] (**14.13**), an anti-HSV drug; Stavudine (2′,3′-didehydro-3π-deoxythymidine) (**14.14**), an anti-HIV drug; and Trifluridine (5-trifluoromethyl-2′-deoxyuridine) (**14.15**), an anti-HSV drug.

The team employed three ionic liquids, $[CH_3OCH_2CH_2mim][CH_3SO_3]$, $[CH_3OCH_2CH_2mim][CF_3CO_2]$ and $[C_4mim][CF_3CO_2]$, as reaction media for some of the key steps in the synthesis of all three antiviral drugs. For example, the first step in the synthesis of Trifluridine involved protection of the two hydroxyl groups of 2′-deoxyuridine (**14.16**) by acetylation to give the diacetoxy product (Scheme 14.3). The reaction in all three ionic liquids resulted in a 90% or more yield of the acetylated product. The next step, trifluoromethylation at the C-5 position, was also carried out in the ionic liquids, resulting in yields of 35–40%.

Overall, the ionic liquids proved to be superior solvents to conventional molecular solvents such as pyridine and DMF. Furthermore, the reactions in the ionic liquids were faster and the amount of solvent required for the reactions was reduced ten-fold.

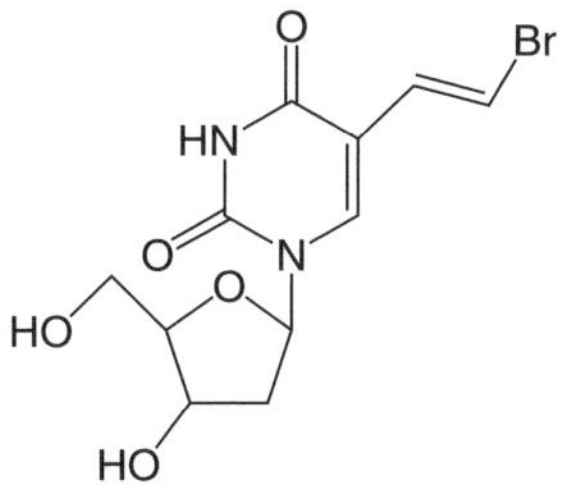

14.13

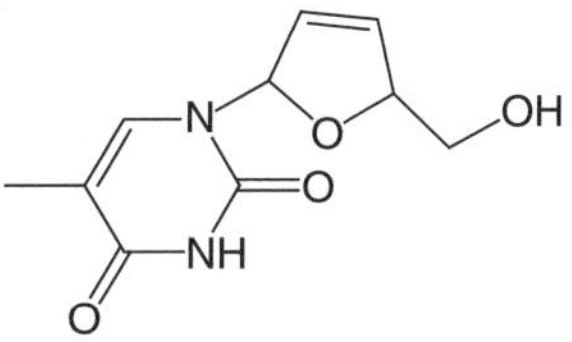

14.14

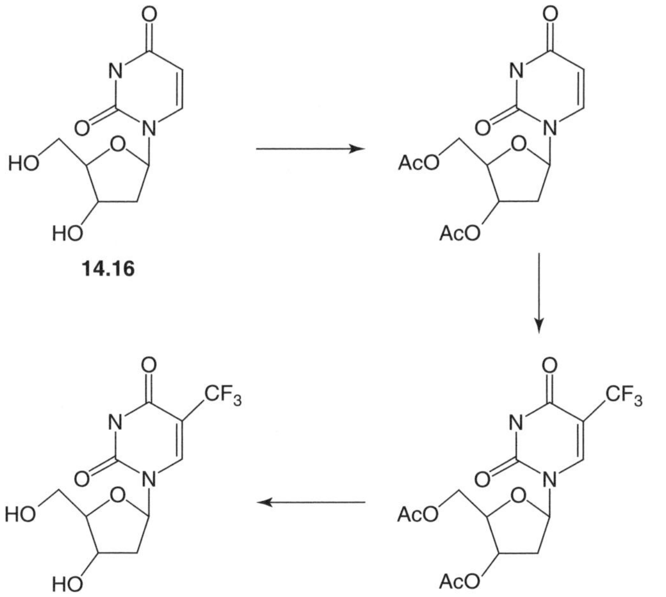

14.16

14.15

Scheme 14.3 Synthesis of Trifluridine.

14.15   EMBALMING AND TISSUE PRESERVATION FLUIDS

Certain ionic liquids have proved to be a good substitute for formalin for embalming and tissue preservation. In 2003, Pernak and co-workers described an investigation into the use of three ionic liquids with 1-alkoxymethyl-3-methylimidazolium cations and $[BF_4]^-$ anions for embalming skeletal muscle from chicken breast and pork and for preserving human skeletal muscle, liver and breast tissue.[27]

Ionic liquids with longer alkoxymethyl substituents in their cations kill bacteria and fungi and therefore prevent the biological decomposition of tissues. The group found that the best embalming and tissue preservation results were obtained with an ionic liquid with an imidazolium cation containing the octyloxymethyl substituent, $[C_8H_{17}OCH_2mim][BF_4]$. This ionic liquid is immiscible with water and has a density similar to that of water.

The animal tissues were embalmed by immersion in the ionic liquid for two years. The tissues did not decay and "kept a pleasant smell." Human tissues fixed in the ionic liquid were mounted on slides and examined by light microscopy. The cell and tissue structures could be analysed with greater precision than similar samples fixed in formalin.

REFERENCES

1. S. N. Falling, S. A. Godleski, J. R. Monnier, G. W. Phillips, J. S. Kanel, 1st International Congress on Ionic Liquids, Salzburg, Austria, June 19–22, 2005, Book of Abstracts, p. 58.
2. M. Freemantle, *Chem. Eng. News*, Aug. 1, 2005, 33.
3. M. Freemantle, *Chem. Eng. News*, March 30, 1998, 32.
4. H. Olivier-Bourbigou and F. Hugues, in *Green Industrial Applications of Ionic Liquids*, ed. R. D. Rogers, K. R. Seddon and S. Volkov, NATO Science Series, Kluwer Academic Publishers, Dordrecht, 2002, p. 67.
5. M. Freemantle, *Chem. Eng. News*, March 31, 2003, 9.
6. D. J. Tempel, P. B. Henderson, J. R. Brzozowski, R. M. Pearlstein, J. J. Hart and D. Tavianni, 1st International Congress on Ionic Liquids, Salzburg, Austria, June 19–22, 2005, Book of Abstracts, p. 35.
7. D. J. Tempel, P. B. Henderson, J. R. Brzozowski, R. M. Pearlstein and H. Cheng, *J. Am. Chem. Soc.*, 2008, **130**, 400.
8. C. Ye, W. Liu, Y. Chen and L. Yu, *Chem. Commun.*, 2001, 2244.
9. C.-M. Jin, C. Ye, B. S. Phillips, J. S. Zabinski, X. Liu, W. Liu and J. M. Shreeve, *J. Mater. Chem.*, 2006, **16**, 1529.

10. Elements: Degussa ScienceNewsletter, 2007, **20**, 26.
11. M. Kömpf, *Linde Technol.*, January 2006, 24.
12. R. J. Singh, R. D. Verma, D. T. Meshri and J. M. Shreeve, *Angew. Chem. Int. Ed.*, 2006, **45**, 3584.
13. H. Gao, C. Ye, R. W. Winter, G. L. Gard, M. E. Sitzmann and J. M. Shreeve, *Eur. J. Inorg. Chem.*, 2006, 3221.
14. C. B. Jones, R. Haiges, T. Schroer and K. O. Christe, *Angew. Chem. Int. Ed.*, 2006, **45**, 4981.
15. A. R. Katritsky, S. Singh, K. Kirichenko, M. Smiglak, J. D. Holbrey, W. M. Reichert, S. K. Spear and R. D. Rogers, *Chem. Eur. J.*, 2006, **12**, 4630.
16. M. Deetlefs, K. R. Seddon and M. Shara, *New. J. Chem.*, 2006, **30**, 317.
17. E. F. Borra, O. Seddiki, R. Angel, D. Eisenstein, P. Hickson, K. R. Seddon and S. P. Worden, *Nature*, 2007, **447**, 979.
18. H. Rodriguez, M. Williams, J. S. Wilkes and R. D. Rogers, *Green Chem.*, 2008, **10**, 501.
19. J. Pernak, K. Sobaskiewicz and I. Mirska, *Green Chem.*, 2003, **5**, 52.
20. K. M. Docherty and C. F. Kulpa Jr, *Green Chem.*, 2005, **7**, 185.
21. J. Pernak and J. Feder-Kubis, *Chem. Eur. J.*, 2005, **11**, 4441.
22. L. Carson, P. K. W. Chau, M. J. Earle, M. A. Gilea, B. F. Gilmore, S. P. Gorman, M. T. McCann and K. R. Seddon, *Green Chem.*, 2009, **11**, 492.
23. W. L. Hough, M. Smiglak, H. Rodríguez, R. P. Swatloski, S. K. Spear, D. T. Daly, J. Pernak, J. E. Grisel, R. D. Carliss, M. D. Soutullo, J. H. Davis, Jr. and R. D. Rogers, *New J. Chem.*, 2007, **31**, 1429.
24. A. K. Prasad, V. Kumar, S. Malhotra, V. T. Ravikumar, Y. S. Sanghvi and V. S. Parmar, *Bioorg. Med. Chem.*, 2005, **13**, 4467.
25. V. Kumar, V. S. Parmar and S. V. Malhotra, *Tetrahedron Lett.*, 2007, **48**, 809.
26. V. Kumar and S. V. Malhotra, *Bioorg. Med. Chem. Lett.*, 2008, **18**, 5640.
27. P. Majewski, A. Pernak, M. Grzymisławski, K. Iwanik and J. Pernak, *Acta Histochem.*, 2003, **105**, 135.

Subject Index